Mr. Wizard's
400
Experiments
in
Science

by

DON HERBERT
TV's MR. WIZARD

and
Hy Ruchlis

BOOK LAB ■ PO Box 206, Ansonia Station ■ New York, NY 10023-0206
1578   ISBN 87594-012-9

How To Use This Book

In the past hundred years we have learned far more about the nature of our world than was discovered in the half million years before that. Why has there been such rapid growth in reliable knowledge in recent times? The increasing use of scientific *experiments is one of the main reasons for this growth.*

Many so-called facts had been believed to be true for thousands of years, yet were not true at all. What is new in recent times is the understanding that "facts" are not really facts until they have been *tested.* And one of the best ways to do that is to design an experiment that tests the truth or falsity of the idea. If the experiment indicates that the idea is false, such evidence becomes far more important than opinions about the facts.

How does one learn to do experiments? By doing many of them, of course. The experiments need not be complicated; in fact, simple ones are best for learning. They need not be original; repeating what others have done before is a good way to learn. The important thing is that the experiments be new to the person doing them and that the person practice observing for themself rather than always taking the word of others for what is supposed to happen.

Something important occurs when you do experiments yourself and make your own observations. You begin to observe things that are not described in the instructions. Sometimes these observations are quite puzzling; often they contradict what you learned before. Then you are in the same situation as a scientist facing the unknown. When this happens to you while doing an experiment, don't drop the puzzle. Face the contradiction squarely. Try to design a new experiment to find out more about the problem and to provide new observations for solving the puzzle.

Don't just sit back and read about the experiments in this book. Be sure to try them yourself. Of course, you will not have time to do all of them, but do as many as you can.

The order in which you do them is not important. This book is organized into short, four-page chapters, each of which deals with a topic. If you are most interested in a topic at the back of the book, simply start there. Skip around as much as you wish. You will find that the chapters are written in such a way that each stands on its own. You will not need any of the information in earlier chapters to understand the experiments in any chapter in the book.

Finally, remember that science books in the library have far more correct information about our world than you can ever possibly obtain by yourself. If you run into a tough problem while experimenting be sure to read some books on the subject. You will find the combination of reading books and doing your own experiments an exciting way to uncover the mysteries of nature.

Don Herbert
Hy Ruchlis

400 Experiments In Science

Table Of Contents

PLANTS

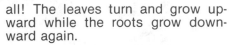

PLANTS

Can you grow a plant upside down? Try it, and see what happens.

Roll a blotter around the inside of a glass. Fasten it with gummed tape. Place several radish seeds near the rim, between the blotter and the glass. Keep the water level in the glass just below the radish seeds for a few days.

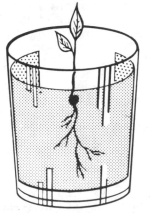

After a few days the seeds sprout. Tiny roots start downward and small stems and leaves go upward.

As soon as the leaves of one radish plant get above the top of the glass, pour out the water, turn the glass upside down, and place the rim on two pencils or strips of wood. Keep the blotter moist by adding water several times a day. Do the plants grow upside down? Not at all! The leaves turn and grow upward while the roots grow downward again.

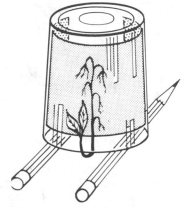

The young plant "senses" that it must send its leaves up and the roots down. Why is it so important for the plant to do this?

Why Do Roots Grow Down?

When you are hungry you go to the refrigerator and get some food. When a plant needs food there is only one way to get it—and that's by making it. In fact the plant has the only real food factory in all the world. Our factories simply start with food that comes from plants and change it a bit to make it more digestible or tasty. The plant is such a good food factory that scientists have not yet been able to make food the way a plant does.

What does a plant need to make its food? You can find out by doing some experiments.

Radish seeds are excellent for these experiments. They sprout quickly and grow well. Plant several seeds to be sure that at least one comes up. If you have no flower pots use a paper cup to plant the seeds. Make a hole in the bottom of the cup and place it in a dish.

It is a good idea to soak the seeds for a day before planting them in earth in the container. This will help the seeds to sprout. Plant your radish seeds about a half inch down in the earth. Water the containers every day.

Grow radish plants for about a week in two different containers. Then stop watering one and keep watering the other. See what happens.

Grow plants under artificial light. A desk lamp may be used. Best results will be obtained with fluorescent lights. Don't let the leaves get closer than about 4 inches from the light.

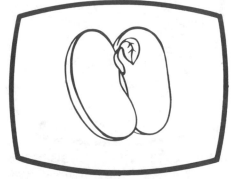

Soak a bean for a day to soften it. Pry apart the two halves of the seed and find the tiny leaves, stem and root of the young plant.

Grow a sweet potato vine. Place the narrow end in water in a jar or glass. Put it in a dark place until roots and stem start. Then put it into the light.

PLANTS

In a week or so, depending upon the size of the containers, the one that gets no water shrivels up and begins to die.

WATER   NO WATER

About ¾ of the body of a plant is water. If water is not supplied, the plant can't make food and it can't grow. And, as the water dries out, the body of the plant shrinks and changes. The plant dies.

You can see this happen with a celery stalk. Cut off the bottom of the stalk and leave it in the open overnight. In the morning the celery is limp. Put the cut end in water. It soon freshens up and becomes firm once again. The water that has been lost is now replaced.

Why must roots grow downward? When it rains, water soaks into the ground. In order for the plant to get water it must send its roots down into the earth where the water is located. Somehow the root senses "down" and grows in that direction.

Why Do Stems Grow Upward?

Try this experiment. Grow radish plants in two small containers. Give both water. But keep one on a sunny windowsill and keep the other in a dark closet.

In a few weeks the plant in the dark closet dies. The one that gets light continues to grow.

Why do plants need light?

The plant makes food from two materials, **carbon dioxide** and water. The carbon dioxide comes from the air. Water comes from the ground. Both are put together inside the leaves of the plants to make **sugar,** the basic food of all living things.

But water and carbon dioxide don't combine to form sugar by themselves. Something must be

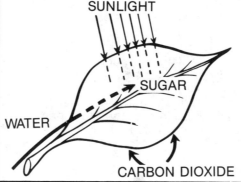

SUNLIGHT

WATER   SUGAR

CARBON DIOXIDE

done to get them to join. Sunlight provides the **energy** which does this important job. This process is called **photosynthesis.**

Try this experiment. Grow a radish plant in the dark until it is about an inch high. Then place it on the windowsill. In a few hours the leaves lean over toward the window as though seeking more light.

Now we can see why the stem of a plant grows upward. When the seed is below ground the leaves must get up above ground quickly to catch light and make food. The seed has some food for the young plant to get started. But this food is soon used up. If the leaves do not come above ground and start working by the time the food in the seed is used up then the plant will die. So it is a matter of life and death for the plant to push its leaves upward.

Wheat, rye, beans, peas, rice, nuts and other seeds make up a large part of our food supply. They are nourishing because the parent plant has stored away a supply of all the important food materials needed by the young plant to get its start in life.

Plants and Minerals

Buy some mineral plant food and make up a solution following the directions on the package. Grow two

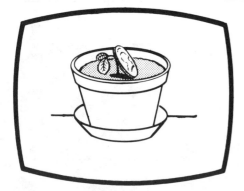

Place a quarter over the spot where you plant a bean seed in a pot. Watch how the young plant lifts the coin and pushes it aside.

ADULT SUPERVISION SUGGESTED

A drop of iodine on moistened starch makes it turn black. Test beans, potatoes, bread and other foods with iodine after wetting the food. Which foods have starch?

Grow 10 plants in one very small pot and 1 in another. The 10 plants crowd each other and do not grow as tall as the one that is alone in the pot.

radish plants in clean sand in different containers. Give one water with minerals. Give the other only water. In a few weeks the one without minerals looks sickly and weak. In a few more weeks it may die.

MINERALS   NO MINERALS

You already know that a plant needs water and carbon dioxide to make sugar, which it uses for food. Part of this sugar is changed into the body of the plant to make it grow. In addition to sugar, the body of the plant needs certain minerals. The most important are: nitrogen, sulphur, phosphorus, calcium, potassium and magnesium.

Water dissolves minerals in the ground and carries them up through the roots and stem to all parts of the plant.

When you burn wood and other plant materials, the ashes that are left are the minerals that were taken in by the water coming up from the ground.

Holes in the Leaves

Try this experiment. Gently rub petroleum jelly (such as Vaseline) on the under sides of the leaves of a young plant. Do the same with the tops of the leaves of another plant of the same kind. Give both of them water, minerals and sunlight.

The plant with the petroleum jelly rubbed on the undersides of its leaves dies. The other one lives.

Why? There are little openings on the under sides of leaves. Carbon dioxide from the air gets into the plant through these tiny holes. If you plug up these openings you stop carbon dioxide from getting in and the plant can't make food. Then it can't live long.

The petroleum jelly on the tops of the leaves doesn't affect the plant because no openings are blocked.

The holes on the under sides of leaves have another purpose. Water travels up the stem from the roots and goes out of the plant through the leaves. If water cannot get out through the holes in the leaves, new water with minerals doesn't come in.

You can see water go up the stem to the leaves of a plant by trying this experiment. Cut off an inch of the bottom of a celery stalk with leaves. Put the cut celery into water colored with ink. In about an hour the leaves and stem are streaked with color.

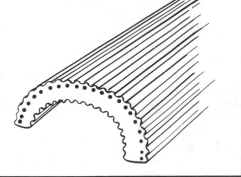

Cut off another small piece from the bottom. Notice the colored spots along the cut end. These are the ends of little tubes through which the water rises in the stalk. Cut the stalk lengthwise and follow the tubes up to the top.

The Importance of Plants

Try this experiment. Put a goldfish into a small fishbowl. In a day or two you will find it coming up for air at the surface of the water. If you leave it alone it will die because of lack of air! Do not let this happen: Add fresh water when you observe the fish coming to surface for air.

It may seem strange to say that there is air in water. But you can see this leaving a glass of cold water on the table. Bubbles of air soon form on the glass as it warms up. These bubbles come from the water.

When a fish uses up the air supply in the water of a small bowl, it must come up to the surface for more air. But this harms the fish and it soon dies.

Put the same goldfish into a bowl of water with some water plants. Now the fish lives much longer. While the plant makes food from carbon dioxide, water, and sunlight, it also forms **oxygen,** most of which it does not need. The fish uses this oxygen to "breathe".

Grow a morning glory plant. Watch how the tip of the stem moves in a circle every few hours. Tie a string to an overhead support and watch the plant wind itself around the string.

Cover a young radish plant with a paper cup that has a hole cut out of the back. Watch the plant seek the light by growing out of the hole.

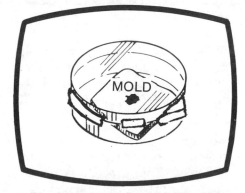

MOLD

Place a piece of moist bread in a closed plastic dish. Seal the cover with tape or glue. Water the mold form. It is a plant. Examine the tiny stems of the mold with a magnifying glass. Do not open the dish after sealing! Give the dish to an adult for disposal.

Fish Help Water Plants?

When a fish burns up food and oxygen it produces carbon dioxide. This goes back into the water. The water plants then use this carbon dioxide to make more food. So you see that water plants and fish help each other, giving what the other needs.

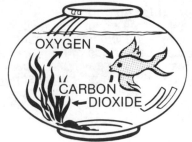

At home the goldfish gets its food from you. But in nature, in a pond or lake, plants are the basic food supply for fish.

The same thing is true of plants and animals on land. Plants not only supply food for animals, they also form the oxygen in the air. And animals use up some of this oxygen to burn up plant food and make carbon dioxide to help new plants grow.

But man makes use of plants in other ways. We use them for making cotton cloth, rope, lumber and paper.

Long ago plants captured sunlight and grew to great size. Today we dig up their remains as coal, oil and gas and use them for fuel and to drive cars, trains, ships and airplanes. When we do we are making use of the energy that plants captured from the sun long ago.

We also use plant materials found in coal, oil, wood and gas to make plastics, chemicals and drugs.

So you see that without plants, civilization would be impossible. In fact we could not remain alive.

Future Food Factories

Scientists are hard at work trying to learn the important secret of the way in which the leaf of a plant makes food from carbon dioxide, water and sunlight. They would like to be able to do this too. Perhaps our food could then be made in factories instead of farms. Food might then become cheaper and more plentiful.

But so far they have not succeeded. This task remains one of the great challenges for the future.

Perhaps you, as a scientist, will some day make an important discovery about plants that will help make this dream of mankind come true.

TRY THESE EXPERIMENTS

1. Plant a carrot. Put the narrow end in water in a jar and watch it grow. Do the same with beets, turnips and parsley.

2. Grow potatoes by cutting off a piece with several buds on it and planting it 3 inches down in a flower pot.

3. Plant lentils, beans, peas and similar seeds that you can get around the kitchen.

4. Grow an onion by setting the bottom in water in a glass.

5. Grow orange and grapefruit seeds in a flower pot.

6. Grow grass seed on a sponge placed in water in a dish.

7. Does a plant need some darkness? Grow a plant under a fluorescent lamp 24 hours a day. It may not grow as well as one that has a regular period of darkness in every 24 hours.

8. Put some mineral plant food solution in a glass. Place it in the sunlight. Algae, a tiny green plant starts to grow after a while.
Pour half of the greenish water into a glass and keep it in the dark. Notice how the one in sunlight becomes much greener.

9. Cut off the leaves of a young plant. See if it grows new leaves.

10. Plant some seeds in a small pot and put it in the refrigerator. Do you think seeds will develop?

11. Try growing pussy willow, begonia, geranium and ivy from cut stems placed in moist sand in a flower pot.

12. Directions for radish seeds call for planting at a depth of ½ inch. Plant some seeds 2 inches deep and see if they come up.

13. Grow a bean plant in darkness and another in bright light. At the beginning the one in darkness grows much taller, in an attempt to find light. After a while, failing to get any it dies.

14. Try to grow a plant on a hot radiator. Does the heat harm the plant?

15. Grow a sweet pea plant. Note the long feelers (tendrils) at the tips of the stems. Rub the under side of a tendril and watch it curl up in a few minutes. The purpose of the tendril is to find support for the plant to climb.

16. Give a growing plant mineral solution that is 5 times as strong as called for in the directions on the package. See how this affects the the plant.

17. Plant stores sell a growth material called "gibberellin". Apply it to several plants according to directions on the package and compare the increase in growth with plants grown in the regular way.

SCIENCE PROJECT

Have you ever noticed the light and dark circular rings of a cut tree trunk? They give information about the life history of the tree.

In the spring and summer the new wood that forms is light in color. In late summer and fall the new growth is darker. This process is repeated each year. Therefore you can tell the age of a tree by counting the number of light or dark rings.

If conditions are poor for many years, the tree grows more slowly and the rings are closer together. This would happen if the tree does not have enough water, minerals or light. Closer rings could be due to overcrowding of trees, disease or insect attack. Irregular or oval shap of rings could be caused if the tree leaned over, or was blocked in growth in some way.

If trees or branches are cut down in your neighborhood, examine the rings. Find out how old that part of the tree is. Can you tell anything about its life history?

YOUR SENSES

YOUR SENSES

Roll up a piece of paper to make a long tube. Sight through the tube at an object. Place your hand alongside the tube as shown in the drawing. You see a big, clean hole right in the middle of your hand!

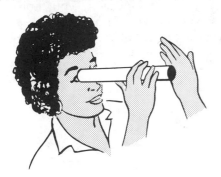

Why do your eyes fool you? To answer this question it will be necessary to learn something about your sense of sight.

Sight

In many ways your eye resembles a camera. Both have lenses that form images. Both have an adjustment at the front for letting in more light, or less. And both have some way of registering the image on the back. In the camera the lights and darks of the images are registered on film. Your eye registers the image by means of numerous **nerves** on the **retina.** Each nerve is like a long electric wire with one end on the retina and the other end deep inside the brain.

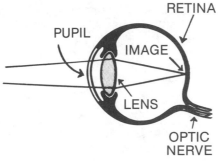

When the sensitive end of a nerve on the retina is struck by light, a kind of electric current travels to a certain spot in the brain. Your brain receives thousands of such electric messages from the many nerves on the retina. Each image on the retina forms a different pattern of such electric messages. Your brain puts together the pattern to recognize the objects outside.

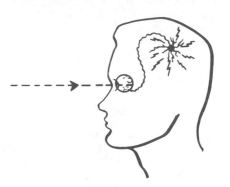

Now look at a window or distant scene. Hold up your finger at arm's length. With one eye closed look at the scene just over the top of your finger. Note the objects that you see Without changing your position

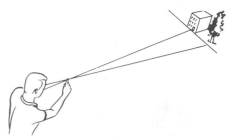

look at the scene with the other eye. The background shifts. You see a different object beyond your finger.

This experiment shows that a different image is observed in each eye.

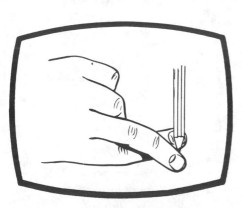

Touch a pencil to the crossed fingers of a blindfolded person. He thinks there are two pencils because he feels them on opposite sides of fingers that are normally in a different position.

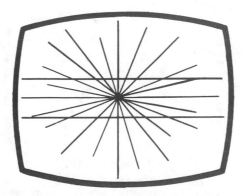

Are the horizontal lines straight or curved? Place a ruler on each line and note that they are straight. Your sense of sight is easily fooled.

Feel a metal surface and a wooden one. The metal feels colder because it conducts heat away from your body faster. It is actually at the same temperature as the wood.

YOUR SENSES

This occurs because each eye is in a slightly different place. Your brain puts together the slightly different images in each eye to form a single impression of the scene. In fact, the slight differences help the brain to judge distance and solidity of objects out in front of you.

You can now see why a hole in your hand was observed when you looked through a tube. The eye that looked through the tube did not see the hand. The eye that saw the hand did not see the round end of the tube. When your brain put the two images together you saw both together at the same time. Thus the hole and the hand appeared in the same place, and you saw a "hole" in your hand.

Touch

Touch the point of a thin nail to different spots on the back of your middle finger, as shown in the drawing. You feel the **pressure** of the nail

point wherever you touch it to the skin. But at some places you feel the point more sharply than at others. These are spots that feel **pain.**

Place the nail on ice for a short while. Then repeat the experiment of touching it to your skin. At some places the nail feels **cold.** At other places you can't feel the cold at all.

Now dip the nail into hot water for a few moments and then touch it to your skin at different places. This time certain spots feel **warm,** while others do not.

Repeat these experiments on the back of your neck. You get similar results, but there are fewer places at which you find the skin sensitive to touch, pain, cold and warmth.

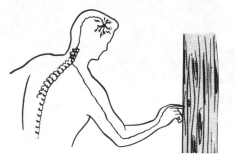

The skin of your body contains thousands of nerves, each of which has a direct electric path to a spot in

the brain. The nerves go up to the brain through the spinal cord inside your backbone.

There are different kinds of nerves. Some are sensitive to touch, others cause a sensation of pain. Still others sense heat or cold. The messages that arrive in the brain from different places all over the body enable you to tell what is happening. Without such a **nervous system** you would not know if a sharp edge is cutting the skin of your leg, or if a finger is being burned or frozen. You would therefore be seriously handicapped in avoiding danger.

A similar system of nerves goes out from the brain to all the muscles and organs of the body. When your brain gets a message that means danger other messages are sent out to the muscles to tell them to move and get you away from the danger. You either pull away your body or use your legs to run away.

Hearing

Make different sounds behind a person's back. Click spoons together. Drop a coin. Scrape a nail on wood. Even though the person cannot see what is happening he can tell that something is moving, and he can often tell what is causing the sound. He can tell these things because of his remarkable

sense of hearing.

Whenever two objects are struck against each other they vibrate rapidly. The air around the objects is set into similar vibration. A sound wave then travels away from the objects through the air. The wave enters your ear and moves into a narrow canal to your eardrum,

Watch people getting off a carnival ride. Some appear to be dizzy after a fast spin on a whirling whip. A liquid in the inner ear was set in motion and continues to spin after the ride was over.

Stare at one spot in a picture for a minute in bright light. Do not look directly at the sun. Then look at a white paper. You will see an **afterimage,** with all colors reversed. This is caused by **fatigue** of the nerve cells on the retina.

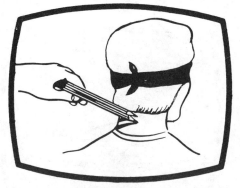

Place the points of two pencils held closely together against the back of a blindfolded person's neck. He feels them as one point. But when touched to his finger he feels two. The sense of touch is much more sensitive in the fingers than on the back of the neck.

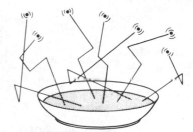

shown in the drawing. The eardrum vibrates. A set of bones are then set into vibration. The vibration finally reaches a coiled-up organ, the **inner ear,** which contains thousands of tiny nerve endings. Different kinds of sounds cause different patterns of electric current in these nerves. These currents race up to the brain, which then recognizes the kind of sound you hear.

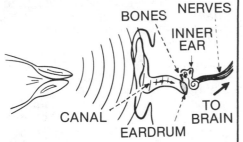

You judge the distance of an object that makes a sound, and its direction, from its loudness and also from the slight differences in the sounds that reach both ears. The sound arrives a tiny bit sooner at the ear nearer to the sound. In addition, the sound in the closer ear is louder. As a result the patterns of electric current sent to your brain from each ear are slightly different. Your brain can judge distance and direction from these slight differences in the electric patterns.

A bat has such highly developed hearing that it can judge from the reflections of its squeaks what obstacles are in the way. It actually catches insects in midair in the dark by using its keen sense of hearing.

what causes evaporation of a liquid such as perfume.

These molecules then bounce around in the air and gradually spread out. Some of them enter your nostrils and affect little nerves, which then send electric messages to the brain. Each substance causes a different pattern of electric messages which tells you what the substance is, or the kind of smell it is producing.

Dogs and many other animals have a sense of smell that is much keener than that of human beings. As a result they are able to recognize individual people and animals just by smell alone. A bloodhound can find a person simply by following the odor hours after a person has walked by.

Taste

Blindfold a friend and feed him a piece of juicy steak while he holds his nose. Ask him what he is eating.

Close one ear with your finger. Ask a friend to click two spoons together anywhere behind you. Try to guess how far away the spoons are and in what direction from you. As your friend moves around you will find it very difficult to tell where he is and how far away.

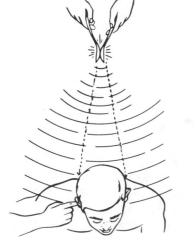

Smell

Place a small amount of perfume in a dish and walk to the opposite end of the room. After a short while you begin to smell the perfume, even though it is far from the dish. Look at the dish. The perfume is gone. It has **evaporated** into the air.

Everything is made of tiny particles called **molecules.** These molecules are always in motion, jiggling back and forth or moving around. Many of the faster molecules of a liquid shoot out into the air and leave the dish. This is

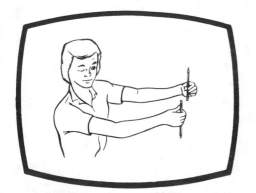

Try to bring the wide end of a pencil down to touch another wide end of a pencil held in front of you. First try it with one eye closed. You miss quite easily. But with two eyes you can do it all the time. Both eyes are needed to sense the distance of an object.

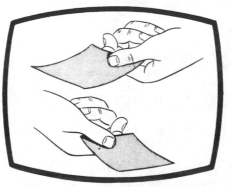

Paste two thin sheets of paper together. Cut out a square of this paper and another of single thickness. Ask a person to tell if they are the same or of different thickness. Most people can tell by feeling them which is thicker. This experiment shows how sensitive our sense of touch is.

Blindfold a person and place any material with an odor nearby. Ask him to tell when it is moved further away. After a few minutes the strength of his sensation decreases and he announces that it has been moved, even though it has remained in the same place.

YOUR SENSES

He is likely to think he is eating paper or stringbeans, rather than delicious steak!

A large part of the flavor of a steak or other tasty food comes from its effect on the sense of smell. If the sense of smell is poor, or if the aroma of a food is prevented from reaching the nose, then wonderfully flavored foods seem to be very ordinary, or even tasteless.

Another part of the taste of a food is due to **taste buds** on the tongue. Dissolve a spoonful of sugar in a quarter glass of water. Dip a toothpick into the sugar solution. Touch it to different parts of the tongue. You find that **sweetness** is detected by means of taste buds mainly at the tip of the tongue.

Try the same experiment with a **sour** material like vinegar. You will find this taste detected mainly at the sides of the tongue.

Try a salt solution. The sense of **saltiness** will be detected mainly at the tip and sides of the tongue.

Try the same experiment with a **bitter** material, such as a bit of onion. First soak the onion in a small amount of water. You will find that the back of the tongue is most sensitive to such bitter tastes.

* * *

There are other senses in addition to those of sight, touch, pain, heat, cold, hearing, smell and taste. For example there is the sense of **balance.** We also have a **kinesthetic** sense in which we are aware of the movement of our muscles and limbs. This sense enables us to tell what the parts of our body are doing at any moment.

The brain puts together the many electric currents arriving from all of our sense organs and thereby gets a picture of what is happening outside and inside of us. This is what makes it possible for us to move about freely, walk or run, drive a car or operate machinery. You can see how important these senses are to our welfare.

TRY THESE EXPERIMENTS

Seat a blindfolded person in a rocking chair. Tilt the chair back a bit. The person can usually tell his position. The position of a liquid in his inner ears enables him to tell his position, even though blindfolded.

* * *

Analyze the taste in foods. First give it to a blindfolded person who holds his nose. Then touch the food to different parts of the tongue. What part of the taste comes from the aroma? What part is due to the fact that it is sweet, salty, sour or bitter?

* * *

Blindfold a person and have him feel different surfaces. In each case ask him to state what the material is and whether he likes the feel or not. Find out whether people differ in their ability to identify materials by touch and what surfaces they like to touch.

* * *

Look in the mirror in a dimly lit room at night and note the size of the **pupil,** the black central spot of the eye. Turn on a very bright light in the room. The pupil closes to keep out an excess of light that might harm the nerves on the retina. Note that it takes a much longer time to return to the original size in the dimly lit room.

* * *

Certain records are sold in which sounds of different pitch are heard. As the pitch rises a point is reached where some people will hear the sound and others won't. Younger people can usually hear higher pitched sounds than older folks.

Look at the magician below, using the right eye, with the paper held about 20 inches away. Slowly bring the paper closer while observing the drawing of the young lady in the corner of your eye. At a certain distance the young lady disappears. With closer approach she reappears. She disappears when her image on your eye falls on the **blind spot,** a region where all the nerves on the retina come together and go up to the brain through the large bundle called the **optic nerve.**

SCIENCE PROJECT

Suppose that a man is driving a car and suddenly sees an animal dash across the road in front of him. Does he step on the brake immediately; or does it take some time for him to respond?

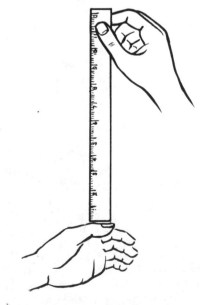

Try this. Hold a ruler at the tip so that it hangs straight down. Ask a friend to put his finger near the bottom of the ruler, ready to grasp it, but not touching it. Tell him to grasp the ruler with his fingers when you let it drop. Drop the ruler. How far does it fall before your friend stops it?

You can tell how much time it took for him to respond by measuring the distance it dropped. Then find the time from this list:

DISTANCE RULER FALLS	TIME IN SECONDS
5 inches	.16
6	.18
7	.19
8	.20
9	.22
10	.23

WATER

WATER

Could you break a bottle with water?

Get a small bottle with a metal screw cap. Fill it to the brim with water. Screw the cap on tightly. Place the bottle in an empty food can. Put the can in the freezer compartment of the refrigerator. Wrap the jar with celophane tape.

In the morning you will find that the glass has shattered inside the can. It was broken by the freezing water.

Why did this happen?

Water is an interesting material in many ways. The bottle broke because of an unusual property of water. Most materials **contract** (become smaller) when they **freeze** or **solidify** (change from liquid to solid). But water does the opposite. It actually **expands** (becomes larger) when it freezes. The force of this expansion is great enough to break iron water pipes on a very cold winter day.

This action by freezing water is very important in nature. It helps to break up rocks on the ground and change it into soil. Water gets into tiny cracks in the rocks when it rains. Then, in the winter, it freezes, expands, and cracks the rock.

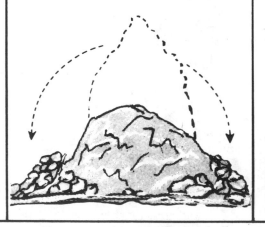

In this manner the rocks on mountains are gradually broken up and worn down. This process of **erosion** can reduce the highest mountains to level plains after many millions of years.

The Force of Moving Water

Water helps erode the land in yet another way. Try this. Get a toy plastic pinwheel of the kind you use at the beach or park. Hold it under the faucet.

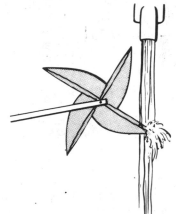

Turn on a gentle stream of water. The pinwheel turns rapidly. Why?

When water moves fast it has a great deal of **inertia** and tends to keep going. It therefore exerts a force against anything that is in the way.

Dampen the back of your hand with a wet cloth. Blow on your hand. The water evaporates and disappears. At the same time your hand feels cool. The evaporation causes the cooling. Your body evaporates the sweat to keep you cool in warm weather.

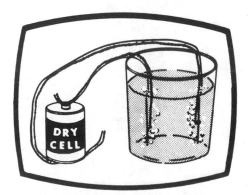

Attach two insulated wires to a flashlight cell and dip the ends into water to which some vinegar has been added. Bubbles are seen to form at each wire. The electric current breaks the water down into the materials of which it is made, **oxygen** and **hydrogen,** which come out in the form of bubbles.

Let some cold water run into a glass from the faucet. Notice the air bubbles. Let the water stand and watch the bubbles of dissolved air come out and float to the top. Some of them cling to the sides of the glass. They resemble small soda water bubbles.

WATER

The faster the water moves, the more force it exerts. The water squirting out of the faucet moves fast enough to turn the pinwheel.

This force of moving water is put to practical use in generating electricity. The water power **turbines** in giant electric power stations are simply larger, stronger, and more efficient "pinwheels" than yours.

A dam blocks the flowing water, which then rises and builds up the pressure below. The water under pressure is permitted to rush through pipes to turn the giant

wheels of the turbine. The electric generators are then forced to turn to make current flow.

Fill a pitcher with water and pour it onto some dry soil. Watch how the water makes a shallow hole in the

ground as the soil is splashed away. Then notice the tiny muddy rivulets that carry some of the soil downhill.

During a flood the force of water may be great enough to carry enormous rocks and boulders for miles downstream. Smaller stones, sand and mud are deposited all along the river edge and in the ocean to form beaches.

In this way the surface of the earth has been changed from bare rock to sand, mud and soil. There are other factors besides water that caused this, but water is probably the most important.

Water as a Solvent

Add a teaspoonful of salt to water in a glass. Stir until it **dissolves** (disappears in the water). Keep dissolving more salt in the same way, a spoonful at a time. Continue until no more dissolves. The **solution** is said to be **saturated.** You will be surprised at the amount of salt

that can be dissolved before saturation is reached.

Try the same thing with sugar and bicarbonate of soda. Large amounts of these materials can also be dissolved.

A liquid that dissolves a solid is called a **solvent.** Water is one of the best solvents known. It dissolves more substances and in greater quantities than the vast majority of other liquids. This property of water is most important in industry, where water is usually used to dissolve substances during manufacture of plastics, paper, chemicals and many other materials. It is used to clean machine parts and automobiles. People make use of the dissolving ability of water to wash and bathe. We make many foods by adding water.

Why is the ocean salty? Rain water trickles through the ground and dissolves some of the solid material, which is then carried down the rivers to the oceans. The accumulation of these dissolved materials through billions of years has made the ocean salty.

Some of the dissolved minerals in the ocean are now being removed and used. For example, one process for making the light metal **magnesium** starts with a mineral obtained from ocean water. This metal is used to make ladders, portable furniture and parts of airplanes.

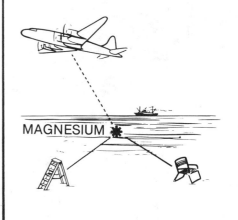

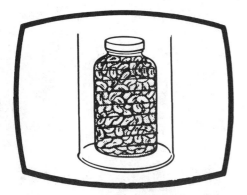

Fill a small jar with beans or peas. Wrap the jar with tape. Then fill it to the brim with water. Cap the jar tightly. Place it in an empty food can. The next day the jar is broken. The water expands the beans and breaks the jar.

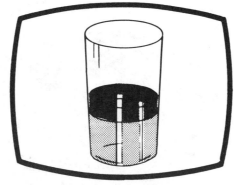

Water does not dissolve everything. Add some cooking oil to water. Mix thoroughly. The oil floats to the top. Add some detergent. The water is now able to break up the oil into tiny bits. That is why we use detergents and soaps to clean greasy objects.

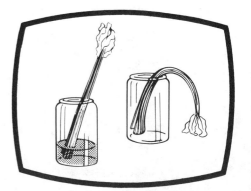

Place a piece of celery in an empty jar. The next morning it is limp because of the loss of water by evaporation. Add some water to the jar. The celery absorbs the water it has lost and stiffens up again.

Heating Water
Caution - Do this with an Adult

Water has a great effect on our lives through its ability to absorb large amounts of heat. You can show this effect by boiling some water in a frying pan.

Select a frying pan with a thick bottom. Put it on the stove. Add a teaspoonful of water. Heat the pan with a low heat. In a few moments the small amount of water begins to boil. This tells us that the temperature of the inside of the frying pan has reached the boiling point of water, 212° F.

Remove the frying pan and let it cool off for several minutes. Then add about ¼ inch of water to the pan. Put it back onto the same low heat. Now the water takes a much longer time to boil than before. In other words it takes a longer time to reach the boiling point of water, 212°F.

Careful measurements show that it takes about 9 times as much heat to raise a certain amount of water to boiling temperature as an equal weight of iron. Therefore water warms up much more slowly.

Let the water boil for a while. Notice how the liquid water boils off in the form of a **gas,** which we call **steam.** Do not let all of the water boil off.

The ability of water to take in large amounts of heat without rising much in temperature explains why you love to go to the ocean or a nearby lake on a hot day. The land warms up much more rapidly than the water of the ocean or lake on a hot summer day. Thus the water remains fairly cool. As a result, the nearby land is kept cool.

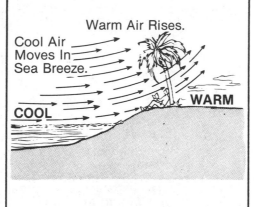

Warm Air Rises.
Cool Air Moves In Sea Breeze.
COOL
WARM

In the winter the water of oceans and lakes cools off more slowly and therefore prevents the surrounding land from getting as cold as it does inland. The climate of continents is changed by this property of water. For example, countries like England, France, Spain, Italy, and Germany, have temperate climates because the warm Gulf Stream carries a tremendous amount of heat to nearby shores from tropical regions. The winds pick up this heat and carry it inland.

In a similar way the western coast of the United States is cooled by an ocean current from Alaska.

Water, Water, Everywhere

You know that the water in the oceans covers ¾ of the earth's surface. But you will also find it in the air and on the land.

Fill a shiny tin can (or thin glass) with ice cubes. Place the can in a dish. Almost immediately the **outside** of the cup becomes moist. Soon water begins to drip down into the dish to form a tiny pool. You will be surprised at the amount of water that can form in this way. Where did the water come from?

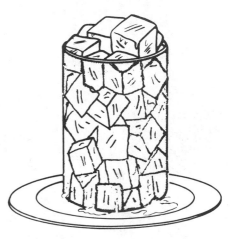

The heat of the sun **evaporates** the water in oceans, lakes and rivers, just as the heat on your stove boils away the water in a pan. In both cases the water remains in the air as a gas, invisible **water vapor.** Later, when the air is cooled the process is reversed, and some of the water comes out of the air in the form of a liquid. That is why water formed on the outside of the tin can. Water forms on cold water pipes and on bathroom mirrors in a similar way.

Hold a metal cup or can over the flame of a candle. You will catch a glimpse of moisture on the outside of the cup. The water comes from the burning of the **hydrogen** in the gas, as it combines with **oxygen** in the air.

DO THIS WITH AN ADULT
Dissolve salt in half a glass of water in a pot, until no more can be dissolved. Heat the water on the stove. You will find that a great deal more salt can now be dissolved. Heat increases the ability of water to dissolve materials.

DO NOT TASTE ANY CHEMICALS
Dry the tip of your tongue with a clean cloth. Place some sugar on the dry spot. You do not taste the sweetness of the sugar. But, a few drops of water dissolve the sugar and you then taste it.

WATER

Can you make a cloud? It's easy. Simply breathe out on a cold day. The invisible water vapor in your breath is cooled, **condensed** and forms visible droplets of a tiny cloud.

A similar cloud forms when you boil water in a kettle. The invisible **steam** is cooled, and condenses to form visible water droplets.

Clouds and rain form in a similar way. When air is cooled enough the water vapor condenses to form large clouds. If the process continues the droplets become large enough to fall. Rain then provides plants and animals with the water they need for life.

A great deal of water exists in the ground. Try this. Collect some earth

from a grassy place, or a forest. Place it in a pot that has been lined with aluminum foil to keep it clean. Heat the pot very gently. You see a mist rising from the soil.

Place the soil out in the open where sunlight can reach it for a few days. Notice how it becomes powdery and dry as the water evaporates. It is the water in soil that the roots of plants soak up to remain alive.

Much of our water supply comes from wells which we dig into the ground. Below a certain level the empty spaces between the grains of soil are filled with water. This water oozes out of the soil to fill up a well that is dug below the ground.

Water and Life

Put some dried beans in a small jar. Put some flour or powdered milk in another. Cover them with water. Put them aside for a few days. Notice how the material begins to change. It starts to smell. Mold may form and grow. Froth may develop. Some of the seeds begin to sprout. After a few days the materials begin to rot.

The dry beans and flour have been standing on the shelf for weeks, months and perhaps years with no noticeable change. But as soon as you add water, living things become active: **abiogenesis.** The seeds begin to develop. Bacteria, molds, and many kinds of germs also develop, because now they have the water they need to live. They feed on the moist beans and flour to cause the rotting. But this does not happen if no water is present.

If you plant the beans in moist soil under the proper conditions they will absorb water from the ground and grow into large bean plants.

Why do plants need water? Examine a carrot. It doesn't look as though it has much water. But grate it into a bowl. Squeeze out the water with a lemon squeezer, or with your hands. Where did all the water come from?

About ¾ of the material of living things is water. Not only is the water needed for the actual building of the bodies of livings things, but it is also the basis of their "transportation systems."

Water dissolves the minerals, gases, body chemicals and wastes and carries them around where needed. Blood does this job in your body. And the water in your blood is what dissolves the other materials and makes it possible for them to reach the different parts of your body, to keep you alive.

SCIENCE PROJECT

How much does water expand when it freezes and becomes ice?

Pour water into a plastic container with sloping sides until it is about ¾ full. Mark the height of the water level on the side of the container. Pour the water into an accurate volume measurer, such as a *cylindrical graduate*. Measure the volume of the water.

Pour the water back into the container and freeze it. The level of the ice will be higher than that of the water. Mark the new level.

Let the ice melt. The water contracts and returns to its old level. Add measured amounts of water until it reaches the former ice level. The extra volume of water added equals the expansion of the water.

By what percentage did the water expand?

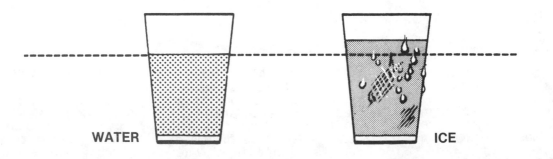

WATER   ICE

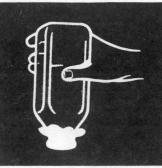

SURFACE TENSION

SURFACE TENSION

Can you make a metal boat float? That's easy if the boat has no holes. But suppose the bottom of the metal boat is full of holes—hundreds of them? Would it then float?

It sounds impossible. But actually, it's very easy to make such a boat. Simply cut a small piece of metal window screening. Bend up the sides to form a small "boat". Place it flat on the surface of water in a dish or pot. The "boat" with hundreds of holes floats on top of the water! Why does this happen? A few additional water tricks will provide some clues.

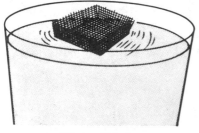

Water Tricks

Fill a glass to the brim with water. Place it on a level surface. Drop nails, clips, washers and other small metal objects into the glass of water, one at a time. You would expect some water to spill out every time an object falls in. But nothing of the kind happens. Instead, the

water rises higher and higher above the top of the glass. You will be amazed at the number of small objects that can be added to the water before it starts to spill over. The level of water in the glass can get to be about ¼ of an inch higher than the glass.

Examine the shape of the water surface. Notice that it is **convex** (bulges outward).

Now spill out some water. Notice that the surface of water is **concave** (curves inward). Observe how the water rises a bit all around the glass, as though it is pulled up by some force.

The convex and concave shapes

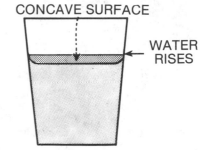

CONCAVE SURFACE

WATER RISES

of the water in a glass make it possible for you to do another interesting trick. Ask your friend to float a small cork in the **center** of a glass of water. Give him the glass and cork and let him get the water. To avoid spilling he is sure to leave some space at the top. When he puts the cork into the water it always moves over to the sides, even if placed in the exact center.

You can easily do this trick by making use of the fact that a floating object will always rise to the hightest position that is possible.

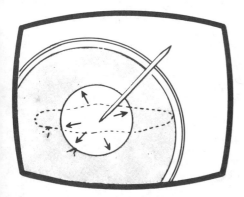

Tie the ends of a 6 inch silk thread to form a loop. Place the loop in water. Touch the inside of the loop with a soapy toothpick. Surface tension is weakened inside the loop and the surface tension outside pulls the thread into a circle.

Sprinkle fine talcum powder (baby powder) on water in a dish. Touch a soapy toothpick to the powder. The surface tension weakens at that point and you see the powdered water suddenly pulled away in all directions by the stronger surface tension elsewhere.

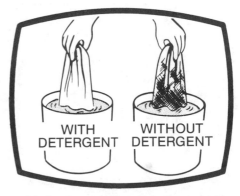

WITH DETERGENT   WITHOUT DETERGENT

Make two cloths dirty by rubbing with grease or oil. Wash one with warm water and detergent. Wash the other with warm water. Molecules of detergent help the water molecules pull the dirt away from the cloth and make it much cleaner.

SURFACE TENSION

Fill the glass with water above the brim. The highest point of the convex surface is then at the center. The cork floats to that position and remains there.

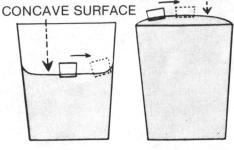

CONCAVE SURFACE

CONVEX SURFACE

On the other hand, when the water level is below the top of the glass, the surface is concave and the center is the lowest point. The cork therefore floats away from the center to the higher places near the sides of the glass.

A Water "Skin"

The fact that the surface of the water can hold up a metal boat with holes is explained with the idea of a kind of "skin" on the surface of the water. Such a "skin" would also help explain why the surface bulges outward when the water is made higher than the sides of the glass and curves inward when the water is lower.

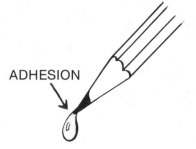

ADHESION

Dip a pencil into water. It comes out wet. In other words water clings to the pencil. The wood **attracts** the water to itself. This force of attraction is called **adhesion.**

Now notice the drop of water that forms on the tip of the pencil. The water seems to stick together to form the drop. There seems to be a force of attraction that keeps the water together. This force is called **cohesion.**

Every material is made up of very tiny bits called molecules. Each molecule of water attracts its neighbors. You see the results of this attraction in the way in which the water bunches together to form drops.

But molecules of water and wood also attract each other. You see the results of this attraction in the adhesion between the pencil and the water that wets it. This attraction explains why the water rises up the sides of the glass to form a concave surface. The molecules of water are pulled up by the attraction of molecules of glass.

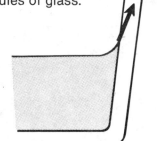

Now let's consider the water alone. The drawing shows a molecule of water (A) beneath the surface of the water. It is being pulled equally by neighboring molecules. Some pull up, others pull down. Some pull to the left, others to the right. The result is that all these forces balance each other and the molecule is rather free to move about inside the water, in any direction at all.

But now consider molecule (B) at the surface. There are no molecules

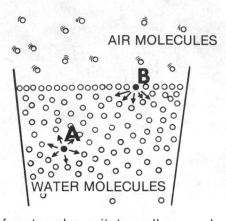

AIR MOLECULES

B

A

WATER MOLECULES

of water above it to pull upward. There are air molecules above, but they are relatively far away and exert little force on the water molecules. On the other hand there are many water molecules below and they pull downward strongly. Therefore the molecules at the surface are pulled downward and are squeezed together. This is what causes the tight "skin" that you observed on top of the water. It is called **surface tension.**

The word "tension" means "tightening force". Surface tension therefore refers to the fact that the surface of a liquid is pulled tightly together, by attractions from inside.

Water is not the only liquid that shows surface tension. In fact, all liquids show the same effect, some more than others.

You can now see why it is possible for your little metal boat to float, despite it holes. It takes some force to break the surface tension of the water. The slight weight of the screen was not enough to break through. So it remained afloat.

Try this experiment. Place the boat on the surface of the water once again.

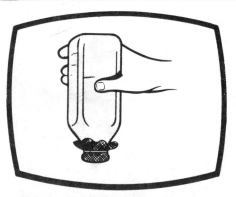

Use a rubber band to fasten a double layer of cheese cloth over the open end of the bottle. Pour in water through the cheese cloth. Invert the bottle. Surface tension and air pressure combine to keep the water from spilling out.

Punch some holes in the side of a can. Let water stream out of the holes. Pinch the streams together with your fingers. Surface tension makes them form one stream.

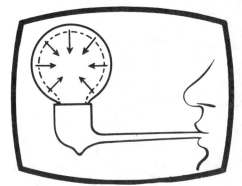

Blow up a bubble. Before it comes off the pipe stop blowing. Surface tension gives the bubble its spherical shape. Watch how it also pulls the bubble together and makes it slowly shrink.

But this time put some kind of weight, such as a penny, inside the boat. Now the boat sinks. It has been made too heavy and breaks through the surface tension of the water.

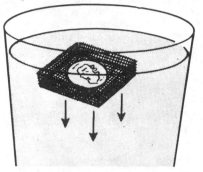

Again put the boat into water. But this time put it in sidewards with the points of the screen piercing the water. Again it sinks. In this case the surface tension was broken by the sharp points and the concentrated weight of the boat on the edge.

Surface tension also helps to explain why you were able to add so many small objects to the full glass of water. The "skin" on the surface of the water acts somewhat like a rubber sheet to keep the water together. Adhesion of the water to the glass attaches this "sheet" to the glass all around the rim. It is therefore possible to build up a height of water above the rim of the glass without spilling over. This makes the upper surface of the water convex.

ACTS LIKE A RUBBER SHEET

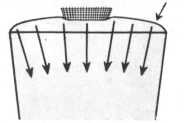

Have you ever seen "water bugs" skimming across the top of a pond or stream? They weigh so little that they can be supported by the surface tension of water under their legs.

Detergents

Surface tension plays a very important part in the action of detergents (including soap) and other cleaning materials.

Float your little screen boat once again. Drop a few bits of detergent powder (or soap flakes) into the boat. In a short time the boat sinks to the bottom.

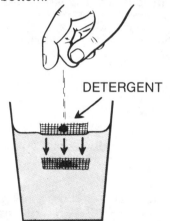

DETERGENT

The detergent weakens the surface tension of the water so that it is no longer able to hold up the little boat.

Try this another way. Make up a solution of soap or detergent and

water. Try floating your boat in this solution. You now fail to do so. You have weakened the surface tension.

This weakening occurs because the molecules of the soap or detergent move between the molecules of water and weaken their attraction for one another. As a result cohesion of the water becomes less and surface tension also becomes less.

Pour some oil on top of a glass of water. Stir it with a spoon to mix the oil and water. You see the oil gather together in round, flat drops on top of the water. Surface tension causes them to pull together. The round shape is the result of the fact that the liquid in each drop is being pulled inward equally from all sides.

Watch the oil. You see the little drops gradually come together and form larger and larger drops.

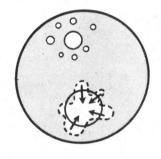

Now add some detergent and mix. This time the oil seems to mix with the water.

When oil is poured on water it has a strong surface tension. So does the water. And the adhesion between the oil and water is slight. So, each pulls toward itself and forms its own bunches. And they do not mix.

But when the detergent is added to the oil and water one end of each molecule of detergent is attracted to

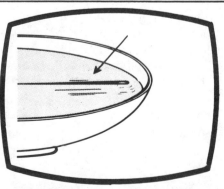

Make "sheets" by letting water from the faucet strike a spoon or knife. Surface tension keeps the water together so as to form the sheet. Many interesting patterns may be made in this way.

Float a needle or razor blade on the surface of water. Gently lower it flat onto the water. Surface tension of the water keeps it afloat.

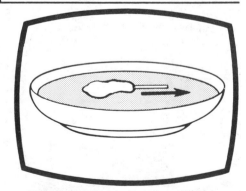

Dip a matchstick or toothpick onto a bar of soap. Place it in water in a sink or bowl. The soap breaks the surface tension and the match is pulled away by the surface tension at the other end.

Try dipping the mater in different household cements and glues. Do these with adult supervision. **19**

SURFACE TENSION

water molecules. The other end is attracted to oil molecules. The detergent molecule thus forms a kind of bridge of attraction between the two. At the same time the surface tension of both oil and water is weakened and they are pulled together by molecules of detergent.

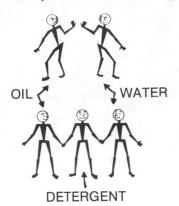

OIL — WATER

DETERGENT

This explains the cleaning action of a detergent. Suppose that your hands are greasy and you wash them with plain water. A film of oil or grease surrounds each bit of dirt. Surface tension of both water and oil keeps the water from pulling the dirt away from your hand.

But with detergent in the water the oil film on the dirt is pulled away and then the water can get at the dirt to wash it away.

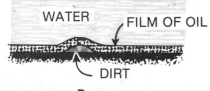

WATER FILM OF OIL

DIRT

Drops

You have already formed drops of oil in water. Now make some drops of water on waxpaper. Use a pencil or medicine dropper to deposit one drop of water on the waxpaper. A round, bulgy drop is formed. Add more small drops of water. They join the first one to form one very large drop. It is round because of the pull

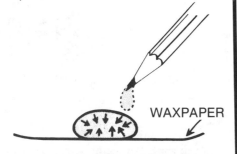

WAXPAPER

of surface tension on all sides. But gravity pulls it down and makes it flatten out.

Now try making an almost perfectly round drop, using oil. Add some oil to water in a glass. Then slowly pour rubbing alcohol down the sides of the glass. The alcohol floats on top of the oil and water. The buoyant effect of the alcohol on the oil tends to cancel out the flattening effect of gravity.

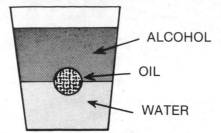

— ALCOHOL

— OIL

— WATER

Now the oil drop appears to be an almost perfect sphere.

Have you ever wondered how glass marbles are made in the shape of small spheres? Blobs of molten glass are dropped from the top of a tall tower. As each blob of glass falls, surface tension pulls the molecules toward the center to give it a spherical shape. By the time the molten glass has reached the bottom of the tower it has hardened into a solid, ball-shaped marble. BB shot is made in the same way from molten lead.

* * *

Scientists have learned a great deal from the study of surface tension. This study has led to important practical results in the manufacture of cleaning materials and chemicals. But equally important is the knowledge about atoms and molecules that have come from this study. This knowledge has been put to use in many fields of science.

TRY THESE EXPERIMENTS

Float two matchsticks in water about one inch apart. A drop of alcohol falling between the match sticks weakens the surface tension and makes the matchsticks fly apart.

* * *

Spray or pour some water onto a dusty screen or floor. Tiny, round droplets form because of surface tension. The dust keeps the water from flattening out and wetting the wood.

* * *

Tie three strings to a can cover. Tie a rubber band to the strings. When you try to pull the cover away from the surface of water the rubber band stretches because of the attraction between the water and the can cover.

SCIENCE PROJECT

The force of surface tension can be measured. You will need a balance device like the one shown in the diagram. You can make one from wood with a pin as a pivot.

Cut a two-inch square of flat plastic. Put a tiny hole in the center with a needle. Tie a knot in a thread and use it to suspend the plastic from one end of the balance. Use very small nails and bits of paper on the other end to balance the plastic.

Bring a pan of water under the plastic until the plastic touches the water. Now add tiny weights (small nails) to the pan on the other side until the plastic suddenly breaks away from the surface of the water. The extra weight required to do that is a measure of the surface tension of the water, the force with which it attracts the plastic.

To find the amount of weight, balance small nails against a nickel. A nickel weighs 5 grams. If 50 nails balance a nickel, then each nail weighs 5 grams, or 1/10 gram.

50

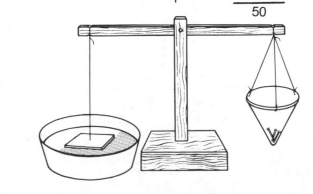

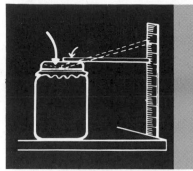

AIR PRESSURE

AIR PRESSURE

Try this experiment. Place a long, thin, wide stick flat on the table, with about 8 inches sticking out over the table edge. The kind of wood used in fruit crates is excellent for this experiment. Cover the part of the stick on the table with two sheets of large size newspaper. Smooth out the paper. Now hit the end of the stick hard with another stick, or a hammer.

You would surely expect the paper to give way. But to your surprise the stick breaks instead. Why does the stick break sooner than the paper?

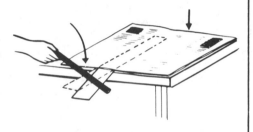

The answer has something to do with air. Let's try some experiments to find out how air helped break the stick.

Air—A Real Material

Would you like to see air? Try this experiment. Tilt an "empty" upsidedown glass under water in a pot, as shown in the drawing. Air bubbles rise to the top of the water. And as the air gets out, water goes into the glass.

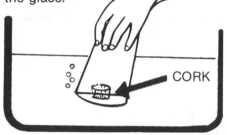

CORK

Actually the glass was not empty at all. It was filled with real material—air.

You can also feel air when it is moving. Notice how the air pushes against you when it is windy. Or, feel the rush of air against your face when you open a vent in a moving car.

You can now see why your stick broke. What seems to be empty space above the newspaper is really filled with air. This air will get out of

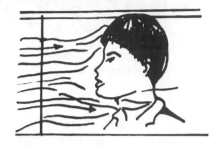

the way if you push it slowly. But when you try to push it out of the way fast, it resists.

When you hit one end of the stick hard, that end went down and the other end tried to move up, like a seesaw. Because of the spread-out paper there was much more air to push than with the piece of wood alone. Thus the air above the newspaper held the wood down and you were able to break the part sticking out over the edge of the table.

Ordinary materials like wood, steel, and bread have weight. Does air have weight? You can show that it does by weighing it with the kind of homemade scale shown in the drawing. Coat hanger wire is good for making this scale.

Adjust the balance by sliding the thread holding the nail, along the wire, back and forth. When the wire is almost horizontal, mark its position along a stick, as shown.

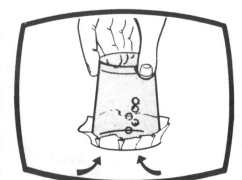

Cover a full glass of water with a waxed paper. Press down along the edges to make a tight seal. When turned upside-down or sidewards, outside air pressure keeps the water from falling out.

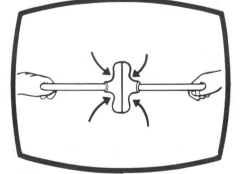

Moisten two "plumber's friends" and push the ends together. Air is pushed out. When the rubber tries to expand a low pressure is caused. Higher air pressure outside keeps the rubber ends together and prevents them from being pulled apart.

Make one small hole in a can of juice. The liquid comes out with difficulty because of outside air pressure. Make another hole and the liquid flows quickly.

AIR PRESSURE

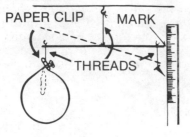

PAPER CLIP   MARK

THREADS

Now let the air out of the balloon by prying the paper clip open. The balloon side of the wire rises slightly, showing that the air you let out had weight.

An Ocean of Air

Could air crush a metal can? Try this.

Get a pint or quart can that has a small screw cap. Clean out the can. heat about half a cup of water in the can, on the hot plate. (Be careful. It's hot). Let the water boil for about a minute. Shut off the heat and quickly cover the can tightly with its cap. Wear safety gloves and carry it to the sink using a stick or pliers. Pour cold water on the can. It collapses as though run over by a truck! Why? **CAUTION:** Wash the can thoroughly with water and a detergent - do not use open flame.

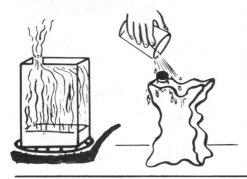

We live at the bottom of an ocean of air, about 100 miles high. The earth's gravity pulls the air down, just as everything else on earth is pulled down. Even though air doesn't weigh much, as compared with other materials, an ocean of air certainly weighs a great deal. The weight of the pile of air above us causes air pressure that is great enough to crush the can.

But why didn't air pressure crush the can without being heated to make steam? The weight of air above us forces the air into every tiny space it can find. With the screw cap removed, air enters the inside of the can and fills it up, pushing outward just as hard as the air outside is pushing inward. So the forces balance each other and nothing happens.

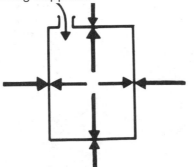

The same thing happens inside your lungs. The air inside pushes outward to balance the push of the air outside. So, no damage is done to your body by the air pressure.

But when you boiled the water, steam pushed out some of the air. You then closed the top and kept the air from returning. As the can was cooled, the steam changed back to water, leaving less air pressure inside the can than before. The greater outside air pressure then crushed the can.

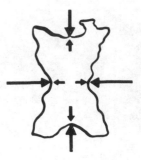

Measuring Air Pressure

Fill a soda or juice bottle with water and turn it upside-down in water in the sink, or in a large pot. You may be surprised that the water does not fall out of the opening at the bottom. It is being held up by air pressure pushing down on the water surface.

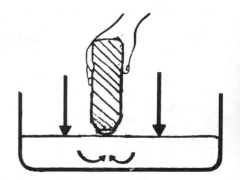

What would happen if the bottle were 50 feet tall, instead of only one foot? This has been tried, except that a long narrow tube was used instead of such a clumsy bottle. Scientists found that air pressure was able to hold up the water in the tube only if it was less than 34 feet tall. When the tube is made taller the water simply remains at the 34 foot level and doesn't go any higher.

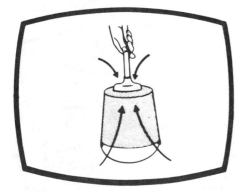

Press a suction cup or "plumber's friend" against a smooth surface. When the rubber springs back it causes a low pressure under the cup and higher air pressure outside makes it stick.

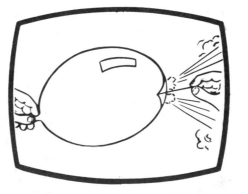

Blow up a balloon. High pressure from your body pushes air into it. Puncture the balloon with a pin. The high pressure causes air to rush out rapidly to make a loud noise.

Put hot water from the sink into a bottle with a small neck. Half fill the bottle with hot water. Attach a balloon to the open end. Heat expands the air, increases the pressure and makes the balloon stand up.

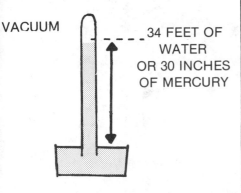

VACUUM — 34 FEET OF WATER OR 30 INCHES OF MERCURY

What's in the empty space at the top of the tube? Nothing! We call such an empty space with nothing in it, a **vacuum.**

When **mercury,** a very heavy liquid, is used instead of water in the tube the liquid goes no higher than about 30 inches.

What is happening? Air pressure on the surface of the liquid is only strong enough to hold up a certain amount of liquid. It holds up a much smaller height of mercury than water because mercury is many times heavier than water.

We have just described a mercury **barometer,** an instrument used to measure air pressure. If the mercury level rises a bit we know that the air pressure has increased, because it can hold up more liquid. If the level drops, the air pressure must be lower because it holds up less liquid.

Scientists refer to the air pressure as a certain number of "inches of mercury". Thus, a weather report might say that the pressure is "29.56 inches of mercury, and rising".

Air pressure is also measured in "pounds per square inch". Normal air pressure is **14.7 pounds per square inch.** This pressure holds up 29.92 inches of mercury.

Changes in air pressure give us an idea as to what kind of weather is coming. Why?

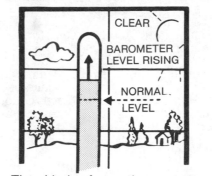

CLEAR
BAROMETER LEVEL RISING
NORMAL LEVEL

The kind of weather we have depends mainly on the kind of air around us. Warm, moist air usually brings cloudy or rainy weather. Cool, dry air usually brings fair weather.

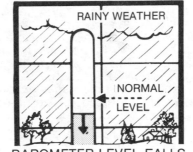

RAINY WEATHER
NORMAL LEVEL
BAROMETER LEVEL FALLS

Since cool, dry air is heavier than warm, moist air; it causes a higher air pressure. Therefore, when the barometer goes up it is a good sign that the weather will improve. And when it goes down, this means that the weather may get worse.

The Aneroid Barometer

You can make a simple **aneroid barometer** (barometer without liquid) in this way. Use a tall glass jar with a wide mouth. Cover the open end with rubber cut from a balloon.

Seal the air in the jar by smearing the outside edge of the jar, and the part of the rubber that touches it, with rubber cement. Wind a string several times around the rubber and tie it to make a tight seal.

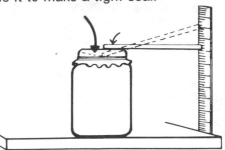

Then cement a long straight broom straw, soda straw or thin wire to the rubber, starting at the center. Mount a vertical, marked stick or ruler on a board, and cement the jar to the board, as shown in the drawing.

Your homemade aneroid barometer is not very accurate, because the rubber goes up and down with changes in temperature, as well as air pressure. This happens because heat causes the air to expand and push the rubber. However, your barometer can still be used for several interesting experiments.

Take the aneroid barometer with you on an auto trip. Watch how the pointer goes up as you ride down a long hill. The lower you go the greater the weight of air above you and the higher the pressure. The higher pressure pushes the rubber inward and causes the opposite end of the pointer to rise.

On the other hand, when the car goes up a hill air pressure becomes less and the pointer moves down.

Try the same experiment in an elevator in a tall building. You get the same results.

Fill a bottle with water through a double layer of cheese cloth covering its top. Turn the bottle upside down in the sink. Air pressure keeps the water from coming out.

Place a book on a balloon and blow into it. High pressure from your body lifts the book. Cars are lifted by air pressure in garages in a similar way.

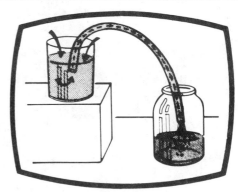

Make a **siphon.** Fill a rubber tube with water. Place one end in a jar full of water and the other in an empty jar. Air pressure pushes water uphill from the full jar to the empty one.

AIR PRESSURE

This type of barometer is used in airplanes to show the altitude according to the changes in air pressure. Such an instrument is called an **altimeter.** The dial barometer used in homes for making weather predictions is an aneroid type.

Instead of flimsy rubber, a practical aneroid barometer uses a flexible metal box that is pushed in by higher air pressure. Air is removed from inside the box to prevent changes due to temperature.

Making Air Move

How can you make high or low air pressure? One simple way is with a **fan.**

Try this. Cut a long strip of paper about 1 inch wide and 11 inches long. Let it hang downward in front of a fan. Your strip bends away from the front of the fan because of a higher air pressure. But when the strip is held behind the fan it bends toward the spinning blades to reveal a low pressure.

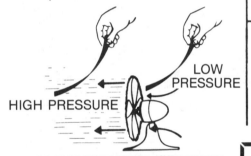

HIGH PRESSURE   LOW PRESSURE

DO THIS WITH ADULT SUPERVISION

The high and low pressures are caused by the action of the blades in pushing air forward. The air is therefore squeezed in front of the fan, to cause the high pressure.

As the air pushed forward, more room for air is left behind the fan, and low pressure is created. A new supply of air then rushes into the low pressure from all sides.

A **vacuum cleaner** is simply a fan with a long tube attached to the low pressure end. The rush of air into the low pressure carries dirt with it. The dirt is deposited in a porous bag at the high pressure end of the fan.

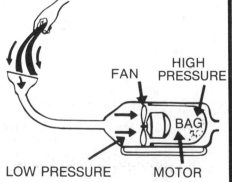

LOW PRESSURE   FAN   HIGH PRESSURE   BAG   MOTOR

Experiment with a vacuum cleaner. Try lifting small bits of paper and thread with it. See what happens to paper streamers. Be sure that you don't use any metal or other solid parts.

There are many uses for air pressure in our modern world. Trains, buses and trucks use high air pressure to operate brakes, and open or close doors. Tires make use of high air pressure. Many water pumps use air pressure to lift water from a well to the inside of the pump. The water is then pushed through pipes.

But the most important air pump of all is in your body.

How We Breath

Feel your **abdomen** (the front part of your body just above the waist). Notice that when you breathe, in your usual relaxed way, the abdomen goes in and out.

When you breathe in, a muscle in your body, the **diaphragm,** moves downward. This creates a larger space inside your lungs than before. The air in your lungs spreads out. The air pressure drops. The higher

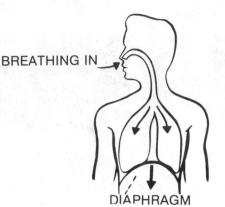

BREATHING IN

DIAPHRAGM

air pressure outside then pushes air into your lungs through the nose or mouth.

You breathe out by pushing the diaphragm up. This squeezes air in the lungs, causes a high air pressure, and your breath rushes out.

In this way the "pump" in your body creates high and low air pressure every moment of your life. Without it you could not live.

You can see that air pressure plays a most important part, not only in our modern world, but also in life itself.

TRY THESE EXPERIMENTS

1. Try a suction cup (or "plumber's friend") against different surfaces. Why does it stick better against smooth surfaces? Try the cup when it is wet. Why does it now work better?

2. Bounce a tennis ball. When it hits the ground the rubber is squeezed and air pressure is increased inside the ball. A moment later the high pressure pushes the ball out again to make it bounce.

SCIENCE PROJECT

Air pressure is due to the weight of air above us. We would therefore expect to find the air pressure to become less the higher up we go.

This is easy to show with an instrument known as an "aneroid barometer" which looks like a flat can-shaped container with a large dial and pointer on the top. If you carry such a barometer up a few flights of stairs you can often detect a slight change in the position of the needle which shows a drop in air pressure.

This drop in pressure is very noticeable when rising in an elevator in a tall building. In some aneroid barometers you can see a slight change by lifting it from the floor to the table.

Take such a barometer along with you on an automobile ride up a mountain. The changes in air pressure are then clearly seen. If you know the height of the mountain you can find out how much the air pressure changes for the change in altitude.

CARBON DIOXIDE

CARBON DIOXIDE

Would you like to make Mr. Wizard's "automatic fountain"? Put a teaspoonful of detergent (the kind that foams) in hot water in a tall jar. Put a lump of **dry ice** inside the jar. A cloud of foam rises to the top of the jar, spills over and creates an amazing fountain that bubbles automatically for a long time.

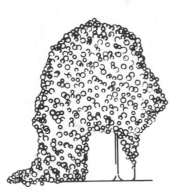

You can obtain the dry ice from an ice cream man, or perhaps at a local store that sells ice cream.

Be sure that you don't touch the dry ice with your bare hands. It is so cold that flesh touched by it is quickly frozen. The frozen flesh feels like a burn.

Handle the dry ice very carefully. Pick up pieces with a spoon or with tongs. If you put dry ice in a metal pot don't touch the cold metal or

you may get a "burn". Keep the dry ice in a carton. If you need small pieces the dry ice may be broken by striking it lightly with a hammer.

Put a piece of dry ice into ordinary water (without detergent) in a bowl or pot. A dense white cloud rises from the surface of the water, spills over the sides, flows down to the table and goes over the edge to the floor.

Observe the rapid bubbling around the dry ice in the bowl. Some kind of gas is being released. It rises to the top. The gas is so cold that

the moist air above the bowl is **condensed** to form a cloud.

Because the cloud is very cold it is heavier than normal air and falls to the table. This is unlike the cloud formed near the spout of a boiling kettle. In that case the heated cloud is warmer than air and therefore rises.

The foam that forms on the top of hot water and detergent is caused by bubbles (gas surrounded by thin films of water-detergent mixture). When the dry ice was put into the jar containing the water and detergent, it changed to a gas and formed bubbles. Since the gas continued to come up from the dry ice, the bubbles were continuously pushed up and out of the bottle. It then looked like an automatic fountain.

Do this with an adult. Light a candle. Let a drop of molten wax fall on the inside of a metal jar cover. Attach the candle to the cover by

Pour some carbon dioxide down a long paper chute toward a lit candle. The chute guides the heavy carbon dioxide gas toward the candle and the flame is put out.

Light a short candle in a tall pot. Pour soda water around the candle. Carbon dioxide from the soda water puts out the flame.

Use dry ice to freeze water. Place a small amount of water in a metal cup and place it on a flat piece of dry ice. The water in the cup soon freezes and forms ice.

CARBON DIOXIDE

pressing its bottom against the molten wax. Put out the flame. Your candle now has a good, safe base for use in experiments.

Make another arrangement like this, but with a very short candle. Place the candles inside a pot, the sides of which are much higher than both candles. Light them. Place a piece of dry ice in the bottom of the pot. In about a minute the shorter candle flame is snuffed out. A while later the tall one is snuffed out.

(Note: Whenever you light matches or do experiments with flames be sure to work over the sink to avoid fires. Keep all flammable materials, such as paper and plastic bags, away from the flame.)

It seems as though the gas coming from the dry ice is heavier than air, sinks to the bottom of the pot and puts out flames. The name of this gas is **carbon dioxide.**

Dry ice is carbon dioxide that has been cooled until it became solid. When you put the very cold, solid carbon dioxide into water it is heated by the water and changes to a gas. This gas bubbles out of the water and goes up into the air. But, because it is normally heavier than air, and is also cold, it tends to fall.

The ability of carbon dioxide gas to put out flames, and its heavieness, make it a very useful fire extinguisher material. One type of fire extinguisher is simply a tank of carbon dioxide under pressure. When the trigger on the tank is pulled the carbon dioxide rushes out and settles down over the flame to smother it.

Making Carbon Dioxide

You can easily make carbon dioxide from two simple household "chemicals". Put a spoonful of bicarbonate of soda in a glass. Pour in a small amount of vinegar. The powder at the bottom of the glass bubbles and forms a froth. While the mixture is bubbling use a pair of tongs to lower a lit match into the glass. The flame goes out.

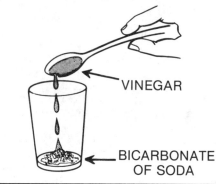

VINEGAR

BICARBONATE OF SODA

Make a new batch of froth in another glass. This time "pour" the gas from the glass over a candle flame. It goes out.

The gas formed by the chemical action of bicarbonate of soda and vinegar is heavier than air and puts out flames. It is the same carbon dioxide that was formed by dry ice.

This method of making carbon dioxide is used in one type of fire extinguisher . When the container is turned upside down an acid mixes with bicarbonate of soda to form carbon dioxide. The pressure of the gas then forces the water and some carbon dioxide over the flame to put it out.

Special Fire extinguishes are used for different fires. All fire extinguishers have use lables. Never use old carbon tetrachloride fire extinguishers.

Another way to get carbon dioxide at home is from soda water. Pour some fresh soda water into a glass.

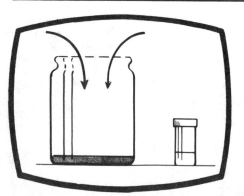

Let limewater stand for several days in an open jar. Put an equal amount in a small closed vial. Slight milkiness of the limewater in the open jar is due to carbon dioxide in the air. The closed vial keeps carbon dioxide away from the limewater.

Put a few drops of vinegar on a piece of chalk. The frothing action is caused by the formation of bubbles of carbon dioxide. A piece of limestone or marble will do the same things. So will baking powder.

Heat a mixture of flour and yeast in an oven. The flour puffs up and forms bread. When cool, break open the bread and notice the many holes caused by carbon dioxide from the yeast. Baking powder may be used instead of yeast.

Place the glass in the sink and lower a lit match into it, above the soda water. The flame goes out.

Soda is made by forcing carbon dioxide into water under high pressure. The bottles of soda are tightly capped to prevent the gas from escaping. When you open the bottle the gas bubbles out and starts to escape.

Add a teaspoonful of sugar to the soda. The bubbles of carbon dioxide come out of the water so fast that it forms a froth. The same thing happens when an ice cream soda is made. In both cases the bubbles come out more rapidly because they have additional surface on which to form.

We refer to drinks made from soda water as **carbonated** because of the carbon dioxide they contain.

Carbon dioxide occurs in other places in your home. But in order to find it we need some way to make it reveal its presence.

A Carbon Dioxide Test

Scientists use **tests** for each substance to discover if it is present in a material. Color, heaviness and appearance, are common tests for solid or liquid materials that we can see. For example, you could recog-nize iron from its metallic appearance, its heaviness and its ability to be attracted by a magnet.

We have seen two tests for carbon dioxide, its heaviness as compared with air, and its ability to put out flames. A third test is to mix the gas with limewater. If the limewater turns milky we have very strong evidence that the gas is carbon dioxide.

Lime is a white powder that you may get in a drug store or from a chemical set. Perhaps your father has some at home. Put a teaspoonful of lime into a glass of warm water. Mix thoroughly. Cover the

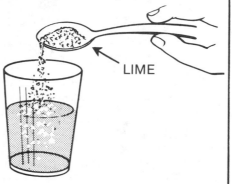

← LIME

glass. Let the solution remain over-night. The next day pour off the clear solution in the upper part of the glass. This is the limewater for your experiments. Keep it in a small, closed jar.

Pour a small amount of limewater into a pint jar. Throw a match into it and shake the jar. The limewater remains clear.

But then lower a lit match into the jar, using a pair of tongs. Wait until the flame goes out. Remove the burned match. Cover the jar and shake it. The limewater becomes very milky.

The carbon dioxide that causes the milkiness of the limewater comes from the burning of the

CARBON DIOXIDE

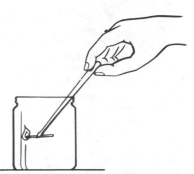

wood. All fuels, such as coal, wood, oil, and gas contain **carbon.** The air around us contains **oxygen.** When the fuel burns the carbon combines with the oxygen and forms **carbon dioxide.**

Since the carbon dioxide is the product that results from burning, it is chemically changed, and therefore will no longer allow burning. When it settles over a flame it pushes away the oxygen that is needed for burning, and the flame goes out.

Carbon Dioxide and Life

Blow through a straw into a small amount of limewater. Before long the limewater turns quite milky. This shows that there is a great deal of carbon dioxide in your breath. Why?

← LIMEWATER

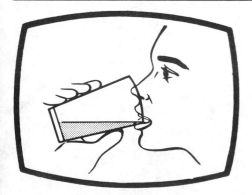

Let a glass of soda stand over-night in the open. In the morning it tastes almost completely "flat". The carbon dioxide has escaped into the air. Place another glass of soda in warm water and note that the gas escapes more rapidly.

Squeeze some lemon juice onto bicarbonate of soda in a glass. The resulting bubbles are carbon dioxide. Any acid will work. Most sour liquids, such as grapefruit juice, are acid and will also do the same thing.

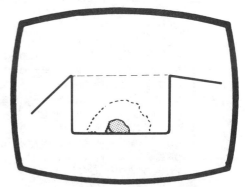

Place some dry ice in an open carton. Note it's size. After an hour the piece is much smaller because most of it has gone off into the air. Unlike ice, the carbon dioxide has skipped the liquid stage and gone directly to a gas. It is therefore "dry" ice.

CARBON DIOXIDE

The food that you eat contains carbon. The air around us contains oxygen. The carbon is slowly "burned" in all parts of your body and combines with oxygen. The resulting energy supplies you with the heat and motion that you need to live. The carbon dioxide that results from the slow "burning" is a waste that is removed from the body by breathing it out.

But perhaps the carbon dioxide that you found in your breath was already in the air that you breathed in? You can test this possibility by forcing ordinary air through limewater. One way to do this is with a rubber syringe. Place the open end of the syringe tube into a small amount of limewater and squeeze the bulb. Air from the bulb

← LIMEWATER

bubbles through the limewater. Remove the syringe and let it expand back to normal shape. It fills with air again. Put the end of the tube in the limewater again and squeeze the air through the lime water.

Do this about 10 times. Little or no milkiness is noticed. Ordinary air seems to have little or no carbon dioxide.

In the previous experiment your breath made the limewater quite milky. Your body somehow added carbon dioxide to the air before you breathed it out.

Does the air around us have any carbon dioxide at all? Squeeze air from the syringe 25 times through a very small amount of limewater, no more than a tablespoonful. You may begin to see a slight milkiness. If not, keep doing it until milkiness is noticed. This experiment shows that a small part of the air is carbon dioxide.

Carbon dioxide is a most important substance. Together with water, it is the basic raw material for life. Plants take in carbon dioxide from the air through holes in the bottoms of their leaves. The roots take in water. The two materials meet inside the leaf where the energy of sunlight makes them combine to form **sugar.** This process, known as **photosynthesis,** is basic to plant life, and is the starting point for production of all of our food.

Plants combine the sugar with minerals taken from the ground to form **proteins.** These proteins then form the bodies and chemical substances of plants. Animals eat plants and use these materials for

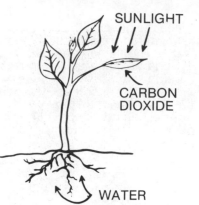

SUNLIGHT

CARBON DIOXIDE

WATER

their own bodies and for the energy needed for life.

Thus, without carbon dioxide our form of life would not be possible.

Scientists think that the amount of carbon dioxide in the air greatly affects the earth's climate. A small increase in the percentage of carbon dioxide in the air seems to cause the earth to keep more of the heat that it gets from the sun. As a result of such an increase the earth has warmed up several degrees during the past 100 years. It is believed that former ice ages and warm periods were caused by the changes in carbon dioxide in the air.

Today, the burning of fuels for power and heat, in factories, homes and cars is causing an increase in carbon dioxide in the air. Many scientists think that this may cause the earth's climate to warm up, causing many changes for life in the future.

SCIENCE PROJECT

Does dry ice have more cooling ability than ordinary ice?

Measure half a cup of water and pour it into a glass. Do the same with another glass. Put a thermometer in each and measure the temperature.

Weigh a piece of dry ice on an accurate scale. (Be sure that you don't touch it directly with any part of your body). Weigh an equal amount of chopped ice. Put the dry ice in one glass of water. Put the ice in the other. Stir the water in each glass gently until all of the solids disappear. Read the thermometers. Which one is at a lower temperature?

If you want to make the measurement more accurate find a way to keep both the dry ice and the regular ice under water while they melt.

DRY ICE   ORDINARY ICE

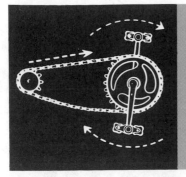

BICYCLES

BICYCLES

Why do you like to ride a bicycle? Try this experiment with a friend on a bicycle. Push him along a level road at a brisk walking pace, while his feet are off the pedals. Note that a small **force** on your part is enough to keep him moving.

Nail a rope to a large, flat board. Try to pull it while a friend sits on it. (You can also use a sled.) You will find it very difficult to make the board budge. A much larger force is required to move him than when he was on the bicycle. Why?

The Wheel

The bicycle makes use of one of man's earliest and most important inventions—the **wheel.** It is in such widespread use today that most of us take it for granted.

Place an empty, round food can with its bottom on the table. Push it. Feel the resistance to your push. This resistance is called **friction.**

There are tiny bumps in both the table and the food can. These bumps tend to prevent the can from moving across the top of the table. As a result you need more force to make the bumps ride over each other.

Place the can on sandpaper. Now it is much more difficult to make it slide. The bumps in the sandpaper

are larger, friction is greater and the force you need to overcome the friction is also greater.

Now try pushing the can while its round side is on the table. Instead of sliding, it now rolls. Instead of the bumps rubbing, they now simply touch and separate. Some force is still needed, but very much less than before.

In the bicycle and other vehicles friction is reduced by use of wheels in contact with the ground. To further reduce friction at the places where the axles touch the wheels, we use **ball bearings** or **roller bearings.** These bearings replace the sliding action at the axles with rolling action.

When you take off the wheels of your bicycle you see these bearings all around the space for the axle. Similar bearings are used to re-

Turn the handle of an eggbeater. Observe how speed is increased by a large gear that turns a small one. The same thing happens with the sprocket wheel of your bicycle, except that a chain is used to transmit your force.

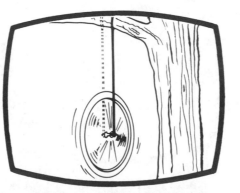

DO THIS WITH AN ADULT
Remove the front wheel of your bicycle from its frame. Support both sides of the axle with ropes and spin the wheel rapidly. Remove one rope from the axle. The spinning wheel does not fall because of gyroscopic action. Instead, it slowly turns around.

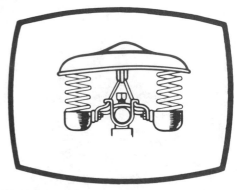

Press downward on the seat of your bicycle. Note how the weight of the rider is supported by two strong springs. They absorb shocks by stretching and allowing more time for the sudden forces to act.

BICYCLES

duce friction under the steering column and at both pedals.

Another method of reducing friction is shown by a simple experiment. Place a small, metal food can on the flat side of a gallon can. Push the small can. Note how much force is needed.

Now put a small amount of lubricating oil on the flat metal can. Push the small can across the oiled surface. It slides easily, almost as though on ice.

The oil gets between the bumps in the two metal surfaces and makes it easier for them to pass over each other. Friction is thereby reduced by the oil. That is why the bearings in your wheels and pedals are oiled or greased. Oil is also used

in automobiles and machinery to reduce the friction of moving parts.

As a result of the use of wheels, ball bearings and lubrication in your bicycle, friction is greatly reduced and it becomes much easier to ride rapidly on a bicycle. You can then travel faster and cover longer distances. That's why you enjoy riding a bike.

Using Friction

Don't get the idea that friction is always a nuisance. There are many places in your bicycle where it is very useful.

The next time it becomes cold enough for ice to form try to walk on it. Observe how your foot slips and it

becomes difficult to walk. The same thing happens to a beginner on roller skates. Or, if your bike goes over an icy spot you are likely to slip and fall over.

One reason for selecting rubber for the tires of bicycles and automobiles is that it is an excellent friction material. It grips the road so that the vehicle can be pushed forward. If it didn't grip the road the

wheel would simply slip and spin without causing the vehicle to move. Ridges and grooves in the tires increase friction on muddy or snowy spots and help prevent slipping.

Operate the brakes on your bicycle while it is in motion. If you have a foot brake the reverse motion of your feet causes a special friction surface at the hub of the back wheel to rub against another surface attached to the wheel. The resulting friction brings the bicycle to a stop.

If you use hand brakes your force causes rubber surfaces to grip the edge of the rolling wheel and friction stops the motion. The brakes of cars make use of friction in a similar way.

The chain of your bicycle also makes use of friction. It would be inconvenient for your feet to reach back to the axle of the rear wheel in order to make it turn. You therefore pedal in the middle of the bicycle and **transmit** your force to the rear wheel. This transmission of force is accomplished by the chain.

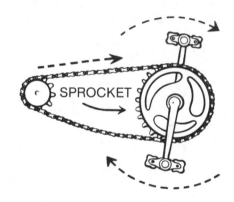

When you rotate the pedals the teeth of the sprocket wheel fit into the spaces in the chain and force it to move. The links in the chain then

Try riding a bicycle with gears. Notice that in low gear you move more slowly, but have enough force to go up a steep hill. You gain force but lose speed. On a level or downhill surface you shift to high gear. You can then afford to lose force in order to gain speed.

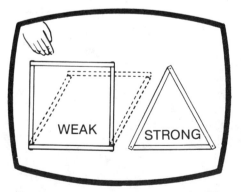

Nail together 4 strips of wood, then three. Compare the strength of each. The three strips are stronger because of the triangular shape. Note how the steel wires of bicycle wheels form triangular shapes near the axle to increase the strength of the wheel.

Make a turn while riding your bike. Notice how you always lean inward to make the turn. **Centrifugal force** tends to throw you outward on a turn. You counteract this force by leaning inward. Airplanes bank on a turn in a similar way. Roads are banked on curves for the same reason.

pull each other and transmit the force to the toothed wheel at the back of the bicycle.

Increasing Speed

Why is the gear at the back wheel so much smaller than the toothed sprocket wheel at the pedals?

Turn the bicycle over onto the handlebars and seat. Note the position of the tire valve on the rear wheel. Rotate the pedals slowly by hand, making exactly one complete turn. Count the number of times that the rear wheel rotates. It will be about three. In other words every turn of the pedal causes the rear wheel to turn much more, about three times.

Turn the bicycle over into its normal riding position. Move the bicycle until the tire valve on the rear wheel is closest to the ground. Make a mark of some kind on the ground near the tire valve.

Now move the bicycle along the ground until the rear wheel has made one complete turn and the tire valve is again closest to the ground. Mark this position. Now measure the distance between marks. It is about 7 feet. This is the distance that your bicycle moves when the rear wheel turns around once.

How far does your bicycle move when you rotate the pedals once? One turn of the pedals makes the rear wheel turn about 3 times. It therefore moves approximately 3 x 7 or 21 feet. Check this by actually riding your bicycle slowly and noting the distance moved in comparison with the turns of the pedals.

DISTANCE MOVED FOR ONE TURN OF PEDALS

A more exact result is obtained by counting the number of teeth in the sprocket wheel and in the toothed wheel at the back, and then comparing them. Try to calculate the distance moved by the bicycle for one turn of the pedals.

The arrangement of toothed wheels and chain greatly multiplies the motion of your feet. Your feet move a rather small distance as they push the pedals around one turn, while the bicycle moves a much greater distance along the ground. This is what makes it possible for you to ride much faster than you can walk. But without a great reduction in friction the effort required to speed up your motion would be greater than you could exert.

Gyroscopic Action

Why is it easier to keep your balance on a bicycle when it is moving fast than when it is moving very slowly?

Try to make a coin stand up on edge. You will probably fail. But give it a push so that it rolls. Now it remains on edge. As it slows down it begins to wobble and finally topples over.

A similar action occurs when your bicycle wheels spin. You can show this in the following way. Turn your bicycle over onto the handlebars and seat. Turn the pedals by hand and make the back wheel spin rapidly. While it is spinning try to tilt the bicycle slightly, sidewards. You feel a resistance to your toppling force. But when the wheel stops spinning you can turn the bicycle over more easily.

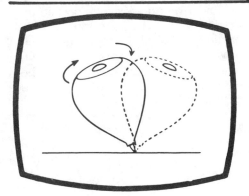

Try to knock over a spinning top. It resists your force and swerves out of the way, while maintaining its upright position. As it slows down it wobbles and finally topples over. These actions are similar to those of the spinning wheels of a bicycle.

Loop a rope over a smooth, sturdy pipe or pulley and pull up a weight. Notice how the rope transmits your force all along its length. And observe that you pull down in order to lift the weight up. The action of this rope resembles that of the chain on your bicycle.

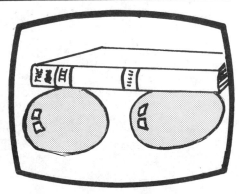

Place a heavy book across two balloons, one of which is at higher pressure than the other. The book sinks deeper into the balloon at lower pressure. A larger area of contact is needed for the lower pressure to support the same weight. A tire at low pressure flattens out for the same reason.

...en a wheel is set into rapid spinning motion it has a great deal of **inertia.** It tends to maintain that motion. When you try to tip over the spinning wheel large forces are required. The inertia of the wheel causes it to resist being made to move in this way. It is easier for the wheel to keep spinning straight, without tipping over or wobbling.

The spinning wheel of a **gyroscope** works in a similar way. Once set spinning it tends to continue in its original position, even in a moving boat or airplane. This provides a way for the pilot to tell how much a ship has turned.

Riding on Air

When you ride your bike do you ever stop to think that you are riding on air? But you know that when the air leaks out of your tire your bike doesn't ride properly. Why?

Puff out the inside of a small plastic bag. Then close the open end with your hand. The trapped air inside the bag now seems to act like an almost solid object. You can support a great deal of weight on the bag without having it give way. But if you let the air out of the bag it collapses.

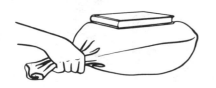

Air is an almost invisible material that takes up space. Unlike solids, it flows and moves out of the way when we walk or ride through it. So, we get the impression that it doesn't exist. However, if it is trapped in a container and squeezed it has no place to go and resists.

Air is **elastic.** It bounces back after the squeezing force is removed. When your tire hits a bump it is suddenly squeezed. The air inside immediately starts to bounce back. But this takes some time. As a result the shock of the bump is smoothed out by the air and your ride is much more comfortable.

Riding on Wires

Notice how your bicycle wheels are made. It would be easy to make a strong wheel out of solid steel. But your bike must be lightweight or you will have to work harder to move.

The bicycle is made much lighter by using thin, lightweight wires for the wheels. How do these wires hold up your weight?

Fasten a thin wire to a small stone. Try to have the stone stay up in the air over the wire. It falls and

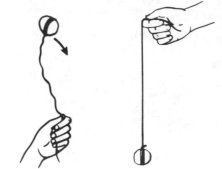

COMPRESSION   TENSION

twists the wire. But when you keep the stone beneath your hand the wire becomes very strong and holds up a great deal of weight.

A wire is very strong when under **tension** (stretching force) because the materials of the wire resist being pulled apart. But when the wire is under compression (squeezing force) it is easy for the wire to give way by bending and twisting without breaking.

Notice how your bicycle wheels are made in such a way that there is always a group of wires in position to be stretched to hold up your weight. As the bicycle wheel rotates, different spokes come into proper position to exert their maximum strength and hold up weight.

We have mentioned only a small number of applications of principles of science in your bicycle. Examine your bicycle carefully. You will find many other ways in which science has been put to work to make it possible for your bicycle to ride smoothly and easily.

SCIENCE PROJECT

It is a simple matter to measure fairly long, straight distances with a bicycle.

Tie a strip of some soft material such as a rubber or plastic tube to the outer rim of one of the wheels of a bicycle. This material must be thick enough to cause a "bump" each time the wheel goes around once.

Hold the bicycle so that this strip touches the ground. Mark that spot on the ground. Move the bicycle straight ahead until the strip returns to the ground and mark that spot. The distance between the marked spots on the ground is the distance traveled by the bicycle during one turn of the wheels.

Now, ride on your bike and count the bumps as the strip touches the ground. Calculate the distance traveled by multiplying the known distance for each rotation of the wheels by the number of bumps you counted.

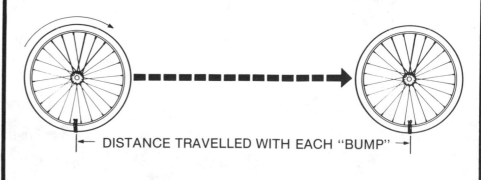

DISTANCE TRAVELLED WITH EACH "BUMP"

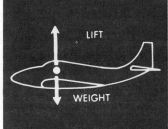

FLYING

FLYING

Place a card across two books of equal thickness, like this:

Can you blow the card off the books? You think so? Try it. You'll be surprised to find it very difficult to do. In fact, you will probably fail, no matter how you blow!

Why does the card act in this strange manner? The answer to this question will help explain how an airplane can fly.

Let's gather some facts with a few more experiments.

Moving Air

Hold a strip of paper in front of your lips, with the strip hanging downward. Blow against the paper. It is pushed away by the air and rises. The moving air creates a high pressure as it strikes the paper.

A kite remains aloft in this way. The wind creates a high pressure against the broad surface of the kite and pushes it up and back.

Now blow over the top of the strip, as shown in the drawing. Again the paper rises. The paper would rise only if the pressure is greater underneath the paper than above it. Since the pressure under the paper is not changed in any way, the moving air must have caused a lower pressure on the top of the paper.

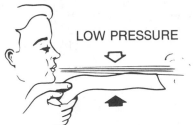

LOW PRESSURE

About 250 years ago, the scientist Bernoulli discovered an important principle that was later used to design airplane wings. According to **Bernoulli's Principle** a fluid (material that flows) has lowest pressure wherever it moves fastest.

For example, the drawing shows a pipe that narrows down at the center. A liquid flowing slowly in the wide part of the pipe must speed up to get by the narrow part, just as the flow of water in the narrow part of a river is more rapid. According to Bernoulli there will be a low pressure in the narrow part of the pipe where the fluid moves the fastest.

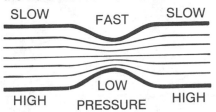

SLOW   FAST   SLOW

LOW

HIGH   PRESSURE   HIGH

Let's see how this principle explains why there was difficulty in blowing a card off two books. When you blow under the card, between the books, the fast moving air causes a reduced pressure. The higher pressure above the card then pushes down and makes it cling to the books.

LOW PRESSURE

On the other hand, when you blow down on the card the air also pushes the card against the books. Thus, whether you blow over or under the card it tends to cling to the books.

TRY THESE EXPERIMENTS

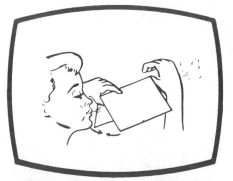

Hold a piece of paper with the edges hanging downward. Blow between the two hanging edges. Fast-moving air causes a low pressure and the two sides of the paper come together.

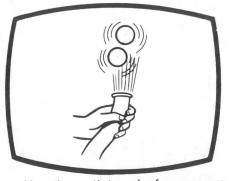

Use the outlet end of a vacuum cleaner to keep ping pong balls and balloons riding on the air stream in midair. The balls are pushed inward into the low pressure of the fast-moving air.

Air blown against a round bottle goes right around it and blows out the candle. This does not happen if a book is used instead of a bottle. The bottle is more streamlined than the book.

Lift

Look at an airplane wing from the edge. You see a shape in which the top of the wing is quite curved. When in flight the position of the wing is set at an angle so that air hits the bottom of the wing. The moving air then creates a high pressure under the wing.

On the upper side of the wing air is forced to speed up a bit as it is jammed into the narrow space between the curve of the wing and the air above it. It therefore moves faster. According to Bernoulli this faster motion creates a low pressure above the wing.

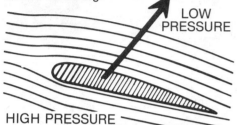

Since the air pressure under the wing is greater than on top, it tends to lift the wing, just as your paper was lifted when you blew across the top or bottom. This upward push of the air is called **lift.** This is what holds up the airplane.

But how can the tremendous weight of a large airplane be supported by the air?

One person couldn't lift up a car by himself. But 20 people could do it.

In the same way the air moving across the top and bottom of an airplane wing can hold up only a small amount of weight on a small amount of wing surface. But if the wing is enlarged, each part adds its bit. All together the small lifts add up to a big one, enough to lift the weight of the airplane. That is why the wing of an airplane is large.

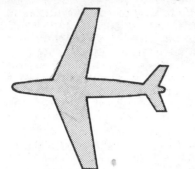

In addition, the lift can be increased by moving faster. The effect of the air in creating lift is then greater. At a certain speed the pressures created by the moving air are enough to raise the airplane into the air.

Moving the Airplane

What makes air move past the wing to create lift? How does an airplane move through the air?

About 250 years ago Sir Isaac Newton explained what made **anything** move. In order to move forward a vehicle must push backward against something.

Blow up a balloon and let go. The action of the air rushing out the back causes a forward reaction that makes the balloon zoom around the room.

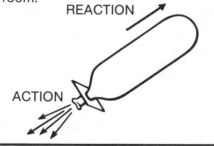

In the same way, the action of the gases moving rapidly backward in a jet airplane causes a reaction that makes the airplane move forward. In a propeller airplane the air is pushed backward by the spinning blades. This backward action on the air causes an equal reaction that makes the airplane move forward. This forward force in a jet or propeller airplane is called **thrust.**

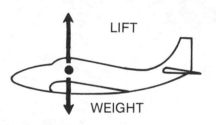

As the airplane begins to move along the ground the wing begins to move against the air and push it out of the way. The motion of the air sliding past the wing begins to cause lift. As the airplane speeds up the motion of air increases, and the lift increases. Finally at the **take-off speed** lift is a bit greater than the weight of the airplane and it begins to rise into the air.

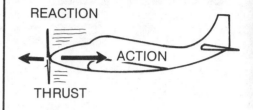

A long runway is needed at an airport to give the airplane enough time to reach its take-off speed. If

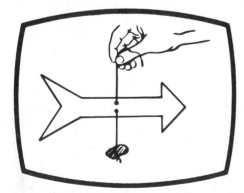

Make a simple weather vane out of cardboard and string. Place it in the wind. The larger drag of the tail forces it back and makes the arrow point into the wind.

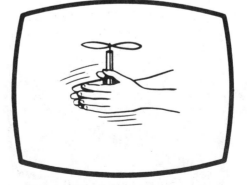

Make a toy helicopter by gluing a small round stick to a plastic or wooden propeller. Twirl the stick in your hands and make it rise into the air.

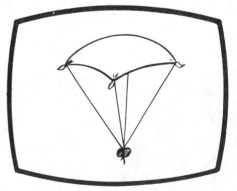

Make a small parachute from thin plastic or cloth. Roll up the parachute and throw it into the air. Increased drag caused by its wide surface slows down the rate of fall.

the airplane is heavily loaded it must get more lift to rise. This requires a higher speed at take-off, and the run along the ground must also be greater.

Drag

You have noticed that fast airplanes usually have rather small wings. With such small wings lift is less and the airplane must move faster before it can take off. It must also land at a higher speed.

With such disadvantages, why are wings of fast airplanes made small? Try the following experiment to find out.

Move a small card against water in the sink or bathtub. First try it with the thin edge of the card moving against the water. Then try it broadside. When the thin edge of the card moves against the water you find that it is much easier to move. There is less **resistance** to the motion because less water is being pushed out of the way.

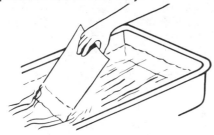

The same thing happens as an airplane moves through the air. The larger the amount of air that must be pushed out of the way by the airplane the larger the resistance of the air. This air resistance in the case of an airplane is called **drag,** because it tends to make the airplane slow down.

The greater the drag the harder the motor must pull to make the air-plane move through the air. Thus, to increase the speed of an airplane designers try to cut down on drag.

One way to reduce drag is to cut down on the size of parts that face the air. Therefore, the wing is made smaller for fast airplanes.

Another way to reduce drag is shown by the following experiment. Cut 3 pieces of card, 1 inch wide and 4 inches long. Fold or roll them up to make the shapes shown in the drawing. Place each on a table about 6 inches from the edge and try blowing against each, one at a time. Which one is pushed backward

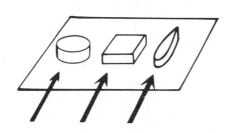

most easily by the air? Which one requires the most force to make it move?

You find that a slight force of air makes the square shape move backward. You have to blow much harder to push the tear-drop shape.

LEAST AIR RESISTANCE

The tear-drop shape has less drag than the square one (or even the round one) because the air flows more smoothly around it. This shape is said to be **streamlined.**

Designers of airplanes use large wind tunnels to experiment with different shapes for parts of airplanes. They seek shapes that reduce drag, and therefore increase the speeds of airplanes.

Controlling Flight

How does a pilot control the flight of an airplane? Let's find out by doing a few experiments.

Make a paper airplane by folding a piece of paper, notebook size, as shown. The heavy paper of a magazine works nicely for such a gliding airplane.

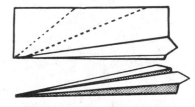

Throw the airplane and let it glide. Notice how it goes. Then turn the back ends up slightly, as shown. Now the glider "noses" upward. If

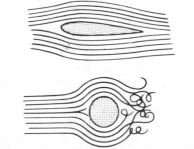

the ends are turned up sharply the glider may nose up until it stalls (stops in midair) and then falls nose downward to the ground, just as a real airplane may stall. If your paper airplane moves fast enough it may make a complete circle (called an "inside loop"). This can be done outdoors unless there is too much breeze.

Attach a cloth or plastic sail to a small boat. When the wind blows, increased drag caused by the wide surface of the cloth or plastic pushes the boat and makes it move.

Place a pin through a card and insert the pin into the hole of a spool. Blow through the hole. The card "sticks" to the spool because of the low pressure caused by fast-moving air.

Hold up a large cardboard when the wind blows. You can feel the large force of drag. Turn the board sidewards and the drag is reduced. Tilt the board and feel the lift.

Turn the back ends of the glider downward. Now the glider's nose hits the ground. Or it may actually land on its back. In an airplane a complete turn of this type in midair is called an "outside loop".

Turn one back end down and the other up. Now the glider rolls over and over, just like the "snap roll" that a real airplane performs.

Your gliding airplane swerves because air striking a turned-up edge pushes against it and thus creates a force that makes it turn.

A real airplane has three main **control** surfaces. The vertical **rudder** at the back of the airplane makes the airplane turn right or left.

The pilot used to control this with two feet by pushing left or right on a pivoted bar. Now all controls are motor driven.

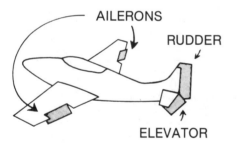

The **elevator**, also at the back of the airplane, moves up or down as the pilot pushes the stick or wheel backward or forward. When the elevator moves down air hits it from the bottom, pushes the tail up, and makes the airplane nose down. When the stick is pulled back the elevator moves up and the tail is pushed down by the moving air. This makes the airplane nose upward.

Finally, there are two **ailerons**, one on the back edge of each wing. When one aileron goes up the other always moves down, and vice versa. These make the airplane lean to one side and "bank".

In making a maneuver more than one control must usually be chang-ed. For example, when making a left turn the pilot moves the rudder to the left and at the same time "banks" the airplane by moving the ailerons.

Rockets

Every airplane has a **ceiling,** or altitude above which it cannot fly. The higher one goes the thinner the air becomes and the less the lift cre-

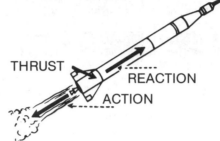

ated by moving air. In addition, the fuels used in airplane engines need oxygen to burn. This oxygen is obtained from the air. Therefore propeller and jet airplanes do not work well at high altitudes where there is little oxygen.

Rockets are designed to operate far above the earth's surface where there is no air. Wings would be useless without such air. Therefore a different method of getting lift is used. A rocket lifts itself straight up by the thrust produced by the rocket engine rather than depending upon the air.

Since oxygen is needed for its engine, a large part of the weight of the rocket is due to the oxygen that is carried along. But then it is free to move about in outer space where there is no air at all.

TRY THESE EXPERIMENTS

1. Put some vinegar into a soda bottle. Wrap some bicarbonate of soda in a tissue and tie it with a rubber band. Insert it into the bottle. Cork the bottle. Place it on two pencils in the sink so that it can roll. Carbon dioxide gas forms from the vinegar and bicarbonate of soda and pressure builds up. The cork pops. Its reaction makes the bottle move backward.

2. Suspend two apples from strings, several inches apart. Blow between them. Bernoulli's Principle causes them to come together.

SCIENCE PROJECT

How much lift do wings of different shape produce? Which shapes are best for airplanes?

The amount of lift can be measured with the arrangement shown below. The wind from a fan must be straightened out before it is useful for measurements. This may be done by passing the air through a set of straight channels as shown at A. A large bundle of tubes, perhaps made of straws, can serve this pur-pose. Be sure to surround the fan with a protector of some kind.

Measure the force of lift on a model of a wing by placing it on one pan of a platform balance. The amount of weight that must be added to that pan to restore balance is a measure of the lift.

Are curved wings better than flat ones? Does it matter at what angle to the wing the wing is held?

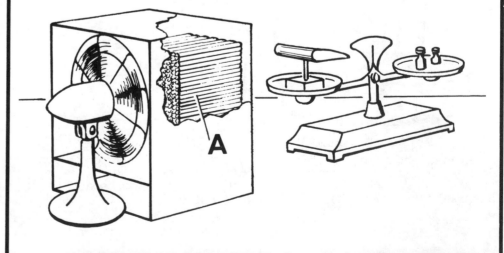

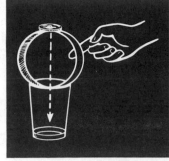

EARTH SATELLITES

EARTH SATELLITES

With a thundering roar the immense rocket slowly rises from the ground. Gathering speed, it quickly zooms up, higher and higher. In a few moments it is gone from view. But it is still gaining speed. Faster and faster it goes.

A short while later, several hundred miles up, a shiny earth satellite streaks its way around and around the earth at the enormous speed of about 18,000 miles an hour, beeping its presence to radio listeners below.

Why does the blast of gas push the rocket up into space? How does it get such immense speed? Why does it stay up there week after week, without any motor?

Let's make our own rocket and find out.

Action and Reaction

You have probably experimented with a blown-up balloon that is released to zoom around the room. You can give this simple rocket a bit more control and make it last longer by putting a cardboard collar around the open end. Experiment with different balloon and collar sizes, to find out which works best. But start with a long, straight type of balloon, rather than a round one. Make the collar as follows.

Cut a one inch square of thin card, from a calling card, or from a file card. Puncture a hole right in the center with the point of a pencil. Push the pencil all the way through to enlarge the hole.

Now push the open end of a long balloon through the hole in the card.

Blow up the balloon as much as you can. Hold the balloon pointing upward while the other hand keeps the air from rushing out. First let go of the open end so that air starts to rush out. An instant later let go of the balloon altogether.

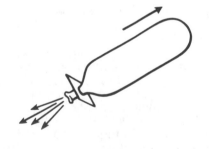

The "rocket" takes off and zooms around the room. Why does this happen?

About 250 years ago a world-famous scientist, Sir Isaac Newton, brought forth a very important idea. **To every action there is an equal and opposite reaction.** This is known as Newton's Third Law of Motion.

When you blow up a balloon the air inside presses equally against all sides of the balloon and therefore

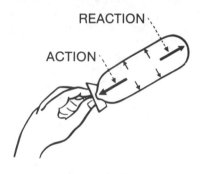

blows it up. The actions on one side are balanced by reactions on the other side. The action forward is balanced by a reaction backward. Since everything is perfectly balanced the balloon doesn't move. It just remains blown up.

Now, as you release the open end, the air rushes out. That's the **action,** in a backward direction. An equal and opposite **reaction** inside the balloon sends the balloon forward.

TRY THESE EXPERIMENTS

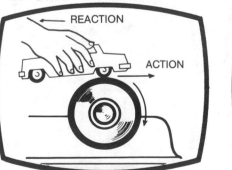

Use a nail to punch a hole in the side of a can near the bottom. Twist the nail sidewards. Suspend the can with strings from a support over a bathtub. When water is poured into the can it comes out sidewards. Reaction makes the can spin the other way.

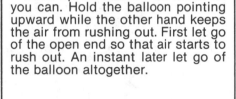

When the back wheel of a toy spring motor car is held against a wheel that is free to move (tricycle or wagon wheel), the large wheel moves backward. This shows that the action of the wheel of a car is in a backward direction.

Wind up the propellor of a rubber band toy airplane. Hold the wheels of the airplane against the floor. Feel the action of the breeze blowing backward against your hand. When you let go the reaction of the air against the propellor sends the airplane forward.

EARTH SATELLITES

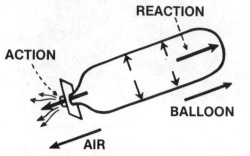

Finally, the air has gone out of the balloon. The action stops, and the reaction also stops. The balloon then falls to the ground.

A real rocket acts in a similar way. Fuel is burned in the rocket. It ex-

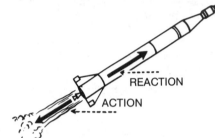

pands and rushes out the back. That's the action. The equal and opposite reaction sends the rocket forward.

If the action is greater, the reaction is also greater. So, if the gases are made to rush out faster, the rocket moves forward faster.

The longer the action lasts, the faster goes the rocket. Thus, the main problem in getting an earth satellite up is to get a fuel that rushes out fast enough, and lasts long enough.

But there is another problem to solve, that of the weight of the rocket.

Four Stage Rocket

Explorer, the first American earth satellite, was sent into orbit on February 1, 1958 with a "four stage" rocket motor. Why was it necessary to use so many "stages"?

Try another experiment with your balloon and card. But this time make a hole in a bigger card, about 3 inches on each side. Blow up the balloon as before and let go.

Now the balloon has difficulty getting off the ground. It is too heavy, and simply slides along the floor. The action of the air rushing out is not great enough to get it up into the air.

We see that it is important to reduce the weight of the rocket as much as possible. But there is a certain amount of weight that we can't avoid. There must be fuel, and something like oxygen to help burn the fuel. There must be some kind of tank to hold the fuel and oxygen. There are pipes and nozzles and other parts of the engine. If all this heavy material remained part of the earth satellite it is doubtful if it could be pushed to the enormous speed of 18,000 miles an hour needed to make it stay in an orbit.

To solve this problem, four separate rocket engines were used in the Explorer, one on top of the other, with the largest at the bottom

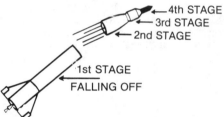

The large first stage engine at the bottom pushed the rest of the rocket up into the air about 50 miles. Then it dropped off to reduce the weight. The smaller second stage rocket then took over and pushed the remaining part higher and faster. Then when its fuel was gone, weight was again reduced by having the second stage engine drop off.

Each stage dropped off as it did its job. Finally, the earth satellite reached its 18,000 mile an hour speed.

Inertia

Why does an earth satellite stay up above the earth without falling down? This brings up another idea that Newton gave us, that of **inertia.**

Set up a pile of checkers. Shoot one checker so as to hit the bottom checker of the pile right in the mid.

dle. The original checker stops short. The bottom checker flies out of the pile just as though it were the checker that you originally used. And the whole pile drops down into the place of the checker that leaves.

Newton explained such events with his idea of **inertia.** According to this idea an object tends to continue whatever motion it has. **An object at rest, tends to remain at rest, and an object in motion tends to remain in motion in a straight line at a steady speed.** Of course, if an unbalanced force is applied the inertia of the object can be overcome. These ideas are part of Newton's First Law of Motion.

The checker that you shoot has **inertia.** It tends to keep moving at the same speed and in a straight line. But the bottom checker is in the way. The motion of the original checker is simply transferred to the second checker which continues with almost the same speed and inertia.

Why does the pile drop down? It tends to remain at rest. The bottom checker shoots out so fast that the

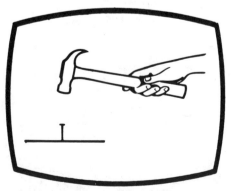

Try hitting a nail into wood using a very light hammer. The nail moves into the wood very slowly. The low inertia of the light hammer makes it easy to move. But at the same time its low inertia makes it have little effect in forcing the nail into wood.

Put a coin on a paper hoop on an open glass. Pull out the hoop suddenly. Inertia leaves the coin in midair. Gravity then pulls it down into the glass.

If you push against a wall while on roller skates. You will roll backwards. Your forward **action** against the wall causes a backward **reaction** that makes you move.

checkers on top have no time to overcome their inertia. So the force of gravity simply pulls them down in the same position to replace the checker that was pushed out.

Now let's use the idea of inertia to explain why the earth satellite does not fall to earth.

Earth Satellite Model

You can make a simple model of an earth satellite using a spool, cord, rubber ball and some small object to serve as a weight. Clay makes a good weight for this purpose. Arrange them like this.

DO THIS OUTDOORS: Make sure that no one is nearby when you do this.

WEIGHT

You can use a small net from an orange bag, or a piece of cloth, to attach the rubber ball to the cord.

Hold the spool in one hand, above your head, and hold the clay with the other hand. Rotate the spool in small circles so as to set the ball whirling over your head. Now gradually let go of the clay in the other hand. If you whirl the ball fast enough it will pull the clay right up to the spool.

Slow down the rotation. At a certain point the clay starts to fall and pulls the ball inward, toward the spool.

Cut the cord holding the clay while the ball is whirling rapidly. The ball shoots away from the spool in a straight line because of its inertia. The cord is pulled up and out of the spool.

INERTIA

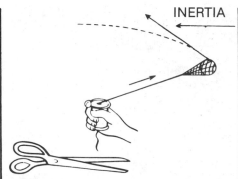

We see that inertia is the cause of the outward pull on the cord as the ball whirls.

What keeps the ball from flying outward? The effect of its inertia is balanced by an inward force. In our experiment, the inward force, known as **centripetal force,** is due to the weight of the clay. In other words, the force of gravity is causing the inward pull.

The outward pull is the reaction to the **centripetal force.** It is sometimes called **centrifugal force.**

When the inward and outward forces are equal, the ball revolves in a circle without flying outward, and without "falling" toward the center of the spool.

An earth satellite stays up for a similar reason. Scientists have figured out that at a speed of about 18,000 miles an hour (5 miles a second) the effect of inertia is exactly

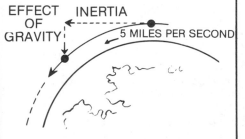

EFFECT OF GRAVITY   INERTIA

5 MILES PER SECOND

right to balance the weight of an object moving parallel to the ground. Inertia at that speed is so great that it carries the earth satellite out into space far enough to just make up for the amount it would fall. So it doesn't get closer to the earth at all.

In fact, if it were to go any faster than 5 miles a second it would actually pull itself away from the earth and take an enlarged oval path. At a speed of 7 miles a second, inertia would be so great that a rocket could actually coast out into space and get away from the earth completely.

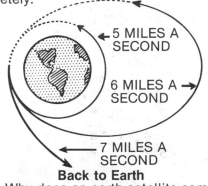

5 MILES A SECOND

6 MILES A SECOND

7 MILES A SECOND

Back to Earth

Why does an earth satellite come back to earth after staying up in space for a while?

Roll a ball along the floor. Its inertia keeps it going until it hits the wall. But roll it on level ground where it has some room to move and you will see it finally come to a stop.

Now roll it on the grass, or on bumpy ground. It comes to a stop very quickly.

Friction slows down the ball. It is caused by the rubbing of one object moving against another. On smooth pavement the ball rolls further because friction is less. On grass or rough ground friction is greater and the ball slows down sooner.

Try this experiment while riding in

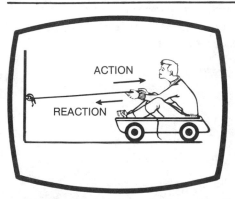

Pull against a rope tied to a wall while sitting in a wagon. As you pull backward the wagon moves forward because of the reaction of the rope and wall.

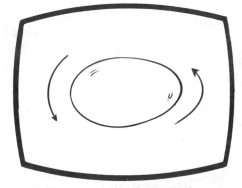

Spin raw and hardboiled eggs. The raw eggs will stop spinning sooner because the loose material inside has more friction.

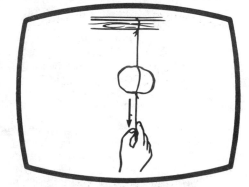

When you pull slowly the upper string breaks because it must support the weight of the rock in addition to your pull. But when you give a sudden pull the lower string breaks. In this case the lower string has more strain than the upper one because of the inertia of the rock.

a car. Open the front vent so that the air blows inward at you. Put your hand in front of the breeze. Notice how it pushes backward at your hand. This backward push, which acts to slow down all moving vehicles is due to inertia of the air. The air tends to remain still. To move through the air you must push it out of the way. The air reacts by pushing back to slow down your motion. A large part of the friction in air is due to this effect of inertia.

Suppose that an earth satellite is sent around the earth at a height of only 25 miles. There is enough air at that height so that the friction of the satellite against the air will cause it to slow down very rapidly. As its speed drops its inertia is no longer great enough to balance the pull of gravity, and down it comes.

But before it hits the ground something else happens. Try this. Rub your palms together tightly. They get very warm. The rubbing of moving parts causes a great ideal of heat.

If an earth satellite had to plow through the air at 5 miles a second friction would cause so much heat that the metal would melt or even burn. The satellite would then be destroyed.

So it is necessary to get the satellite up out of the air before it can be free to go round and round the earth.

But even 150 miles up there is a tiny bit of air. Friction with this small amount of air gradually makes it come closer to the earth. Then it gets into slightly thicker air. Friction increases. Finally, it falls back to earth, glowing brightly like a meteor.

Why Go Up There?

Mankind lives on the earth. It gives him food, shelter, clothing, materials for industry, and energy.

He wants to make this earth a better place in which to live. Strange as it may seem those little earth satellites can give a great deal of information to improve our lives.

They help us predict what the weather will be like in the future. We learn more about planets and stars from earth. They enable us to make more accurate maps of the earth to help navigators.

And, of course, we are also taking the first steps for space travel. We now know what the other side of the moon looks like. Men have landed on the moon.

And like every great scientific achievement, we never can really foresee what will come from our first earth satellites. They are likely to lead to new and wonderful discoveries.

Of course there are also dangers in space, such as wars in space. It will be up to us to make sure wars do not occur and that space flight will continue to benefit humanity.

TRY THESE EXPERIMENTS

1. Throw a ball hard while sitting in a swing. Reaction makes the swing move backward.

Move your legs forward on the swing. The swing moves in the opposite direction.

2. Do this experiment outdoors. Put water in a tin can. Hold the can by the rim and whirl it around in a vertical circle, at arm's length. The water does not come out at the top of its path because of its inertia in the same way that an earth satellite stays up because of its inertia.

3. Put a marble inside a round-bottomed bowl. Make the marble roll along the upper rim of the bowl by moving the bowl in small circles. Inertia of the marble keeps it from falling to the bottom of the bowl.

4. Sit in the back seat of a rowboat with the front end touching a dock. As you move forward in the boat, reaction makes the boat move backward. When you arrive at the front end the boat is quite far from the dock.

5. Hold a garden hose lightly in one hand while a friend suddenly turns on the water, full force. Feel the backward reaction to the forward action of the water.

If you place the hose on the ground, reaction makes it snake back and forth.

SCIENCE PROJECT

As an earth satellite moves around the earth it is seen from the ground in different places at different times. The path is not simple because the earth rotates while the earth satellite revolves. But scientists can predict where the satellite will appear. How do they do it? You can work out a simple case as follows.

First, make a model of the actual path of the satellite. Do this by cutting out a circular hole in a large cardboard that is just big enough to fit around a globe model of the earth. Make 12 marks evenly around the circle. Fit the card over the globe so it stays in the same position while the globe rotates.

Now suppose that the satellite revolves around the earth in six hours. Each of the 12 marks on the circle therefore shows where the satellite is located every half hour. The globe usually shows 24 half circles from North Pole to South Pole. Each circle shows how much the earth rotates in one hour.

Start at any of the 12 points on the cutout. Turn the globe to show a half hour of rotation. Move to the next mark on the cutout. This shows on the globe where the satellite will appear a half hour later. Mark that spot. Keep doing this one mark at a time, and trace the path of the satellite on the globe.

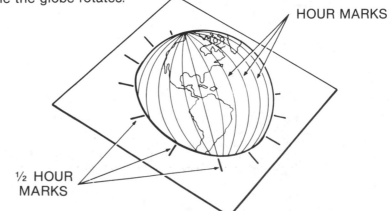

HOUR MARKS

½ HOUR MARKS

CENTER OF GRAVITY

CENTER OF GRAVITY

Stick a needle into half a small potato. Then challenge your friend to make the potato balance freely on the needle point. He will probably fail.

Show him how to do it. Stick two long, fairly heavy forks into the potato as shown in the drawing. With a few adjustments of the forks, the potato may be balanced quite neatly on top of a soda bottle.

Why does the potato balance in this way? A few experiments will help you understand the answer to this question.

Finding the Center of Gravity

Can you predict in advance the exact point on a card at which it will balance perfectly when placed on the point of a nail?

Cut a piece of cardboard, with the kind of irregular shape shown in the drawing. Punch a series of holes anywhere in the cardboard, using a nail.

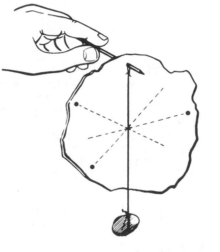

Now push a smaller nail through one of the holes. Let the cardboard swing freely. It soon comes to rest. Locate the **vertical** (downward) line from the **pivot** (position of the nail) by letting some kind of weight hang down from it, using a string. Let the cardboard, weight and string swing freely until they come to a stop. Use a pencil mark to locate the position of the string on the cardboard. Draw the line showing the position of the string. This is the vertical direction from the pivot.

Remove the nail and repeat the process from the other holes. You will be surprised to find that all the lines meet in a single point!

Now balance the cardboard on the point of a nail as shown in the drawing. When you place the point of the nail right under the point on the card where the lines meet it can be made to balance. But at every other point the cardboard topples.

The point of balance that you have found is called the **center of gravity.** You might think of it as the "center of weight".

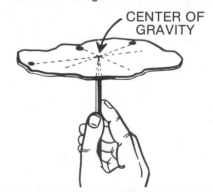

CENTER OF GRAVITY

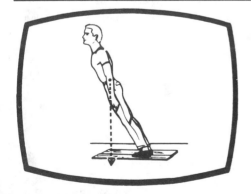

How far forward can you lean without toppling? Use screws to attach a pair of **discarded** shoes to a board. You can then use these shoes with an enlarged base to lean far forward without toppling.

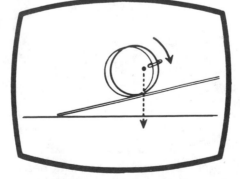

The paper hoop rolls uphill a short distance because the weight of the clip shifts the center of gravity to one side. The center of gravity actually moves downward as the hoop rolls uphill.

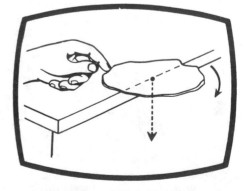

Locate the center of gravity of an irregular, flat object by pushing it slowly over the edge of a table. At the moment that the center of gravity passes the table edge the object begins to topple.

CENTER OF GRAVITY

Now try this experiment. Punch a hole with a nail at a point near the center of gravity of the cardboard. Insert a nail into the hole. Try to get the cardboard to balance with its center of gravity above the nail. No matter what you do the cardboard rolls around and the center of gravity always ends up directly below the nail.

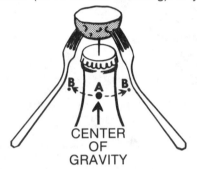

CENTER OF GRAVITY

Now push the cardboard to one side. Notice that this raises the center of gravity. Let go. The cardboard swings back and forth like a pendulum until the center of gravity ends up at the lowest point it can reach.

You see that the center of gravity seems to act as though all the weight of the cardboard were at that point. **Scientists have found that in dealing with the earth's pull of gravity on an object they can think of the entire weight as though it were concentrated at one point, the center of gravity.**

Why did your method of finding the center of gravity of the cardboard work out correctly? When the cardboard is allowed to swing freely from any point it always comes to rest with its center of gravity directly below the pivot. Then when you draw a vertical line (downward), the center of gravity is somewhere on that line.

When you do this again from a different point, the second line also contains the center of gravity. Since both lines contain the center of gravity, the place where they meet locates its position.

Center of Gravity in Midair

Find the center of gravity of a wire coat hanger. Let it hang freely from a nail as shown in the drawing. Repeat from several different places on the coat hanger. The location of the center of gravity seems to be somewhere inside the triangle, in midair.

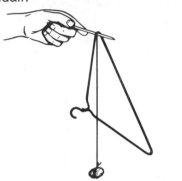

You can find the location of the center of gravity by clipping a sheet of paper to the triangle and marking off the vertical lines, just as you did with the cardboard.

What happens to the position of the center of gravity if you fasten a weight to one side of the coat hanger? Try it. Attach a fork to the

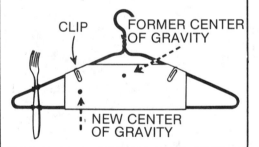

CLIP   FORMER CENTER OF GRAVITY

NEW CENTER OF GRAVITY

coat hanger as shown in the drawing. Locate the new center of gravity. You find that the extra weight causes the center of gravity to shift closer to the fork. You see that the position of the center of gravity depends upon the locations of the different parts of the object.

Stable or Unstable?

Why did the potato balance on the point of a needle when the forks were stuck into it?

With the heavy forks hanging far down, the center of gravity of the combination of potato, needle and forks is shifted below the point of the needle, somewhere inside the bottle (at A in the drawing). If you

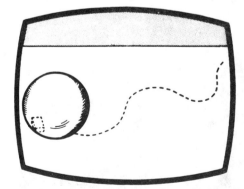

CENTER OF GRAVITY

push the potato slightly to make it topple, this causes the center of gravity to be raised (from A to B). The earth's force of gravity then tends to pull the center of gravity down again. Therefore the combination of potato, needle and forks does not topple when pushed.

It is said to be **stable** and tends to return to its original position when pushed. It is stable because the center of gravity is below the point of support, the tip of the needle.

When the forks are removed the center of gravity of the potato and needle is shifted upward, above the

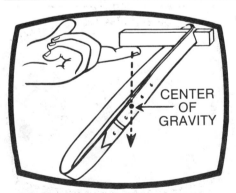

CENTER OF GRAVITY

Can you balance a piece of wood with one end on your finger? Cut a slanted notch in the wood and use a belt to shift the center of gravity inward.

Ask a friend to carry a heavy weight. Notice how he leans to the opposite side to bring the shifted center of gravity over his feet to keep from toppling.

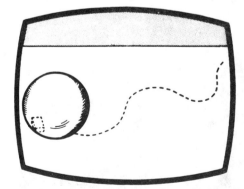

Make a ping pong ball wobble as it rolls by inserting a metal object inside the ball. The peculiar wobble is caused by a center of gravity that is shifted off-center.

point of support. Now the potato is in an **unstable** position. The center of gravity can be pulled downward to a lower position by the force of gravity and the potato topples.

Try this experiment. Place a ball on a level table. It remains where you place it. Change the position of the ball in any way you please. It remains in the new position. If you give it a push it doesn't topple, but simply rolls a bit and comes to rest.

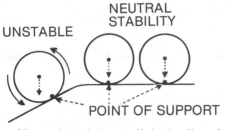

NEUTRAL
STABILITY

UNSTABLE

POINT OF SUPPORT

Now place it on a slight incline. It rolls downhill and doesn't stop until it reaches the lowest point.

On the level table the ball illustrates a condition of **neutral** stability by taking any position in which you place it. The center of gravity of the ball is right at its center. As a result, on a level surface the center of gravity is always exactly over the point of support, as shown in the drawing. Therefore it can't be pulled any lower and keeps its position.

But on a slight incline the center of gravity is no longer directly above the point of support and can be pulled to lower positions. So the ball rolls downhill.

Your Stability

Play a trick on your friend. Ask him to pick up a handkerchief placed near his feet on the floor, without bending his knees or moving his feet. Most young people will be able to do this.

But now have him stand with the backs of both shoes touching the wall. This time he can't pick up the handkerchief, no matter how hard he tries.

A few experiments with a narrow, empty box will explain this trick.

Place the box on its narrow end. It is stable and remains in that position. The empty box has a center of gravity that is practically at its center. A vertical line from that center of gravity (C in the drawing) falls between the ends of the box at A and B. The line AB is said to be the **base**, which supports the box.

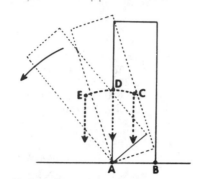

If you push the box to make it topple, it pivots at one of its ends (A). This causes the center of gravity to rise, from C toward D. When you let go, after pushing the box slightly, the center of gravity is pulled back to C, and the box returns to its original position. It is stable.

Now push it a greater distance. Keep pushing, slowly. At a certain position of the center of gravity (D) the box begins to topple over. At this position the center of gravity is at its highest point, directly over the pivot. Beyond that position (at E) the box is unstable and falls.

This experiment shows that so long as the vertical line from the center of gravity falls inside the supporting base an object is stable. If the vertical line from the center of

gravity falls outside the base then the object is unstable and topples.

You can now see why it is impossible to pick up a handkerchief from the floor when the backs of your shoes are against the wall.

A person has a center of gravity, like any other "object". This center

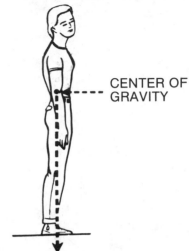

CENTER OF
GRAVITY

of gravity is located near the waist. When you stand up straight this center of gravity is above your feet. A vertical line from the center of gravity falls inside your supporting base, your feet, as shown in the drawing.

But when your shoes are against the wall and you lean forward, the

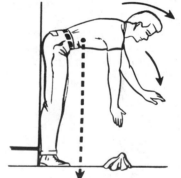

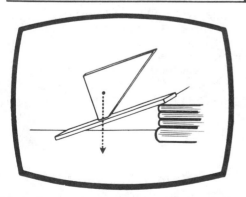

When placed on its small edge the block topples. If the board on which it rests is tilted sufficiently the block does not topple because the center of gravity then falls inside the base.

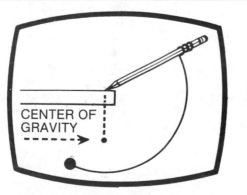

CENTER OF
GRAVITY

Can you balance a pencil on its point? Attach a weight and wire so that the center of gravity of the combination is under the pencil point.

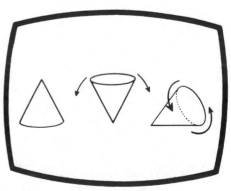

A cone-shaped paper cup is stable on its wide base and unstable on its tip. When on its side it shows **neutral** stability and rolls when given a slight push.

CENTER OF GRAVITY

center of gravity also moves forward. Soon, the center of gravity is too far forward. A vertical line from the center of gravity is now outside your supporting base. Over you go!

Now watch your friend as he picks up the handkerchief while standing in the middle of the room. Notice how his legs actually lean backward at an angle. This causes his center of gravity to shift backwards by the same amount that his bending body makes it move forward. As a result, the center of gravity remains over his feet and he does not topple. His body is now in a stable position.

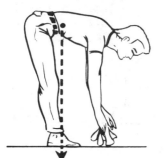

INCREASING STABILITY

Now try this experiment. Have your friend stand with his feet close together. Give him a slight push from one side. Your friend finds it necessary to shift his foot on the side opposite the push, in order to keep from falling over.

Now try the same thing with his feet spread wide apart. A slight push seems to have no effect on him. Even a big push may not cause him to budge.

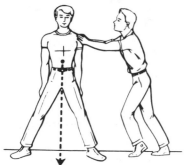

This experiment shows that stabiltiy is increased by **widening the base.**

This happens because a larger base requires that the center of gravity move a much bigger distance and be lifted through a greater height before it moves out beyond the supporting base. This fact explains why people tend to stand with feet wide apart while on moving buses and trains.

Why are your feet large, compared with the hooves of a horse or the paws of a large animal? A four-legged animal has such a wide base that it has little difficulty in standing, walking, or running. The young of such animals usually walk at birth or a short time after birth. But a human being has a much smaller base when walking upright. A large-sized foot increases the base and therefore helps in walking upright. Even then, it takes infants about a year to learn how to walk.

Lower Center of Gravity

There is another reason why people have greater difficulty in learning to walk than animals. Try the following experiment.

Put a heavy stone inside the bottom of the narrow box that you used in a previous experiment. Now you have to push the box a great deal further before it topples. The extra weight lowered the center of gravity, and the box became more stable.

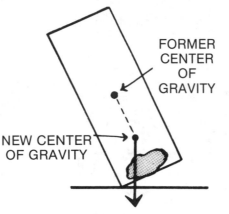

FORMER CENTER OF GRAVITY

NEW CENTER OF GRAVITY

It is harder for you to learn to walk than for a four legged animal, not only because your base is smaller, but also because your center of gravity is higher. You are therefore less stable than the four legged animal.

Place the box (used in the previous experiment) on its side. Now it is impossible to make it topple. Its base is as wide as possible, and its center of gravity is as low as possible. Its stability is therefore the greatest possible for that particular box.

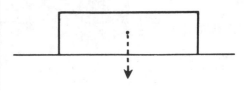

The great stability of a four-legged animal resembles that of the box in a low position. On the other hand, when you stand up, a higher center of gravity and a small base give you poor stability, like that of the tall narrow box on its end. But you then have the advantage that two limbs are freed for grasping objects. This is what makes it possible for us to use tools, make things, and even to write.

Engineers make use of these facts in designing vehicles and machines. For example, one reason why cars are made low and wide is that a low center of gravity and a wide base make it more difficult to topple over.

You can see that center of gravity and its effect on balance play a most important part in our lives.

SCIENCE PROJECT

You know that the moon revolves around the earth. But did you know that the earth revolves around the moon—at least a tiny bit?

Attach a large ball of clay to one end of a thin metal rod, and a smaller ball at the other end. (A thin coat hanger is excellent for this purpose). The balls of clay represent the earth and moon. Now suspend both of them by a single thread attached to the wire. Where do you have to tie the cord so that the rod remains balanced—in other words, in a level position? This balance point is the center of gravity of the system.

Now spin the earth-moon model gently. Both the "earth" and "moon" revolve around the common center of gravity, the balance point where the cord is tied.

In the real earth-moon system the common center of gravity is inside the earth, but about 3000 miles from the earth's center. The earth actually revolves around this center of gravity about once a month, and so does the moon.

Find out how to locate the center of gravity if you know the weights of the large ball and the small one.

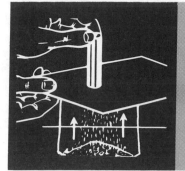

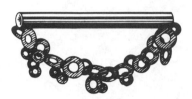

MAGNETISM

M A G N E T I S M

Look at a magnet. Do you see anything around it? Move your hand through the air near the magnet. Do you feel anything?

You see and feel nothing. Yet there is "something" around the magnet. Let's prove it.

You may have several strong magnets at home. If not, you can buy then at a local hardware store. **Alnico** magnets will be best for your experiments.

You will also need some iron filings (tiny bits of iron). You can make enough for your experiments by steel wood with a scissors. Do this in a plastic bag with a "zip" lock. The filings will not scatter. Caution: When working with filings protect your eyes. Do not rub then with your fingers.

Now place a magnet flat on the table. Shake up the filings in the zip lock bag. Do you see the beautiful pattern formed by the filings? We call this pattern a **"magnetic field."**

This "something" is not even like air, which can be trapped in a bottle, and has weight. The magnetic field around a magnet has no weight at all. It passes right through the plastic of the bag.

Repeat the experiment by placing a sheet of aluminum foil, or a piece of cardboard, or plastic or even several more sheets of plastic between the magnet and the iron filings. The bits of iron form the pattern just as though these materials were not there. We see that magnetism can go right through solid objects.

Magnetic Poles

Bring the head of a small nail just underneath the middle of a rod-shaped alnico magnet, and let go. The nail tends to jump over to the end of the magnet and stick there.

Dip a magnet into a pile of small iron objects, such as washers or brads. They stick mainly to the ends of the magnet. A tiny bridge of iron objects may form from one end of the magnet to the other.

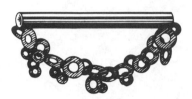

These experiments show that the ends of the magnet have greater magnetic pull than the middle. The places of greatest magnetic strength in a magnet are called **magnetic poles.** In a rod-shaped magnet there are two magnetic poles, one at each end.

A magnet may have more than two poles. But no magnet has ever been found to have just one pole.

North and South

Suspend your magnet from a thread so that it can rotate freely. Keep it away from iron objects, such as pipes and radiators. See that no other magnets are nearby. Your magnet swings around and points in one particular direction. Mark one end with a permanent marker or nail polish.

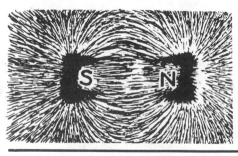

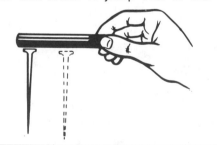

T R Y   T H E S E   E X P E R I M E N T S

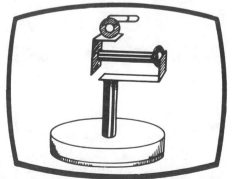

Make a magnetic "building" using several strong magnets and various iron or steel objects, such as washers, nails, small steel plates, and parts cut from food cans.

Place a pile of books between a compass and an alnico magnet held in your hand. The magnetism goes through the books as though they weren't there, and makes the compass needle move.

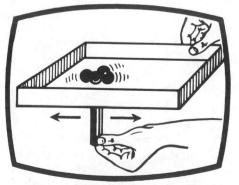

Place a number of steel balls from an old skate wheel in a cardboard box or plastic dish. The balls will "roll mysteriously" as you move an alnico magnet under the box.

MAGNETISM

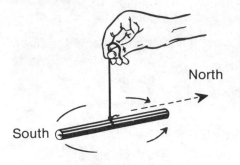

Try to change the direction in which it points. When you let go it immediately comes back to the same direction.

Place it in different parts of the room—away from iron objects. Take it outdoors. No matter what you do the magnet swings back to the same direction.

Locate north in your room. You can use a magnetic compass for this purpose. Or you can obtain north from the direction of the sun's shadow at noon. You can also use a map.

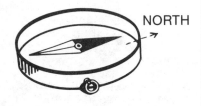

Your hanging magnet points in almost the same direction as north. The end of the magnet that points north is called a "north seeking pole". Most people shorten this name to **north pole.**

The opposite end of the magnet points south. So it is called a **south pole.**

Your swinging magnet is almost the same as a regular magnetic compass. The main difference is in the way the magnet swings. A regular compass has a magnet that pivots on a sharp point and has a case around it.

Two Kinds of Force

Locate the north pole of an alnico magnet by suspending it from a string. Use a crayon to mark the pole that points north. Do the same with another alnico magnet.

Now hold a magnet in each hand with the north poles facing each other. Bring them together. They almost wriggle out of your grasp as they resist being pushed together. The magnets **repel** each other.

REPULSION

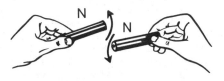

Try bringing the south poles together. They repel in the same way. Don't try this too often, because the magnets tend to get weaker as you repeat the experiment.

Now try bringing a north pole near a south pole. This time they **atttract** and pull together tightly. Try pulling then apart. You can get some idea of the strength of the magnets by the amount of force you must exert to separate them.

ATTRACTIION

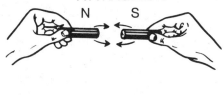

These facts are expressed in the following rules:
Like poles repel.
Unlike poles attract.

You can quickly check the truth of these rules by bringing the north end of your magnet near a compass. In most compasses the north seeking end is darker in color than the south end. The north end of the compass swings away from the north pole of your magnet, and the south end comes closer. The reverse happens with the south pole.

Why does the magnetic needle of a compass point north? The earth behaves as a giant magnet with two magnetic poles. One magnetic pole appears to be in Northern Canada. Compasses therefore point towards this region.

Magnetic Materials

Bring your magnet near paper, wood, glass, plastic, a copper penny, aluminum foil and a rubber band. Nothing seems to happen. These are **non-magnetic** materials.

But bring the magnet near a "paper clip", steel wool, nails, washers, a tin can, a radiator. Now there is attraction between the magnet and the object. All of these objects are made of iron. And iron is the most common **magnetic material.**

Certain other metals, (nickel and cobalt) are magnetic, but not as much as iron. Certain **alloys** (mixtures of metals) are also magnetic.

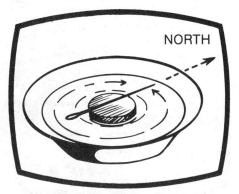

Magnetize a needle by touching its tip to a south pole of an alnico magnet. When placed on a floating cork, the needle is turned around by the earth's magnetism and points north.

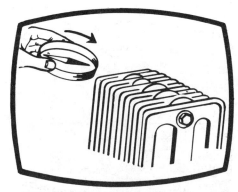

Test a radiator with a compass to see if it is magnetized. The top is a south pole and the bottom is north. It has been magnetized by the magnetic field of the earth.

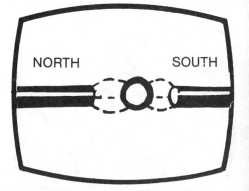

Place a steel washer between north and south magnetic poles. Then make an iron filings pattern. Notice how the lines of force seem to crowd into the washers.

For example, **alnico** is an alloy of **al**uminum, **ni**ckel, **co**balt and iron. When non-magnetic aluminum is alloyed with the magnetic materials, nickel, cobalt and iron, it makes one of the strongest magnetic materials known.

ALuminum
NIckel
CObalt
Iron
} ALNICO

Not everything that contains iron is magnetic. Try lifting a stainless steel spoon with a magnet. In most cases no attraction will be noticed, even though the spoon is made mainly of iron.

Magnetic Theory

Why are some materials magnetic while others are not?

Scientists think that the atoms (tiny particles) of all materials are little magnets, each with a north and south pole. If the atoms are arranged with their poles helter-skelter, as shown in the drawing, then the attraction or repulsion of a north pole is cancelled by the opposite force of a nearby south pole. So, magnetic force is not noticed outside the magnet.

ATOMS DISORGANIZED

NOT A MAGNET

But if most of the atoms have their poles lined up as shown in the drawing then the attraction or repul-

ATOMS LINED UP

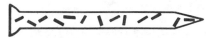

A MAGNET

sion of all the north poles at one end are not cancelled out, and the magnetic force is therefore noticed. At the other end of the magnet the forces exerted by the south poles of the atoms build up into the strong force of a south pole.

The center of a magnet is weak because the force of every north pole is cancelled by the force of a nearby south pole.

Making a Magnet

Long ago certain rocks were found to attract iron. These rocks were the first magnets. Soon, a way was found to make magnets starting with these natural rock magnets. How was this done?

Pick out a nail about 1 or 2 inches long. Try to pick up a smaller nail (or iron washer) with the larger nail. You will probably not succeed. But touch or rub the nail against the pole of a magnet and it thereafter picks up several small nails by magnetic attraction.

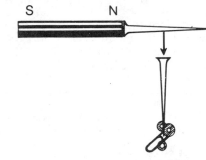

Bang the larger nail against a hard object a few times and try to pick up the small nails. The magnetism of the nail has become much weaker. What happened?

Materials differ in the ease with which their atoms can be lined up. In non-magnetic materials the atoms don't turn around and line up. But in a magnetic material like iron they line up very easily. So, when an

iron nail is brought near the north pole of a magnet, the south poles of the atoms of the nail are pulled around and line up facing the north pole. Thus, the nail becomes magnetized.

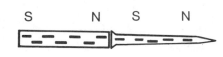

But if the nail is taken away from the magnet, its atoms get out of order by themselves and then the magnetism is not noticed outside the nail. Actually, it is there all the time, but in a jumbled condition. Banging the magnet helps to shift the atoms so that they lose their lined-up condition.

Lines of Force

In the first experiment, on page 45 you saw how iron filings can be made to reveal the magnetic field around a magnet. Why did the iron filings line up to form a pattern?

As each filing fell, the poles of its atoms were attracted or repelled by the nearby poles of the magnet under the paper. So each iron filing became a tiny magnet. The north pole of one iron filing then attracted the south pole of another. Thus, all the filings tended to stick together to form a curved line around the magnet, from the north to the south pole. The paths along which the filings form are called **lines of force.**

Electromagnets

About 150 years ago a way was found to make a magnet using electric current. You can make such a magnet as follows.

Obtain a length of thin, insulated magnet wire (about 20 feet of #30).

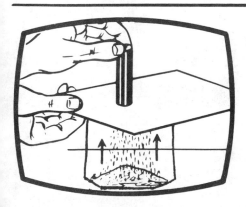

Add some salt to the iron filings. How can you get them apart? Wrap your magnet in plastic and place it in the bag. When you lift the magnet, the salt remains.

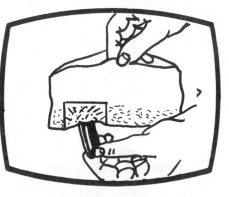

Place an index card in your filings bag with most of the filing on top of it. Hold the magnet under the card in the bag.

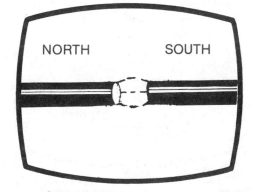

NORTH   SOUTH

Make a lines of force pattern with unlike poles near each other, then with like poles. Lines of force always go from one pole to an unlike pole, and never to a like pole.

MAGNETISM

Such wire may be purchased in electrical and hardware stores. Wind it round and round a nail about 2 inches long. Leave 6 inches of wire free at both ends, for making electrical connections. Remove the insulation from about one inch of each free end of the wire. If the insulation is a brown enamel scrape it off using a dull knife or the edge of a screwdriver. Be sure that the clean copper wire shows through all around the wire at both ends.

Touch the ends of the wires to a flashlight cell (battery). Electric current flows. Now the coil and nail can pick up small iron objects. When you pull the wire away from the flashlight cell the current stops and the iron objects drop off.

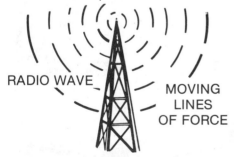

FLASHLIGHT CELL

A coil of wire wound in this way is called an **electromagnet.** It will work without the iron in the center. But the iron center or **core** makes the electromagnet much stronger.

Using Magnetism

With such an electromagnet you can make a distant iron object move by simply turning current on and off. This fact makes the electromagnet useful in telegraph sets, electric bells, telephone receivers, and electric motors.

In fact, all of radio and television is based upon electromagnetism. A radio wave is made up, in part, of a moving magnetic field, in which the lines of force spread out as a wave.

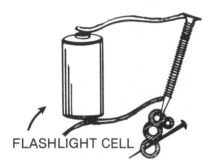

RADIO WAVE — MOVING LINES OF FORCE

Scientists therefore refer to a radio wave as an "electromagnetic wave".

The light which enables you to see is also an electromagnetic wave. In fact, radio waves were discovered because an English scientist named Maxwell developed a theory that light was an electromagnetic wave. Other scientists, looking for different kinds of electromagnetic waves then discovered radio waves.

You can see that magnetism plays a very important part in your life. It is all around you, in the form of light, radio, TV and the many electrical devices that you use every day.

TRY THESE EXPERIMENTS

1. Does an electromagnet have poles? Bring each end near a compass while the current is on. You will find that the electromagnet has a north and south pole just like a regular rod-shaped magnet.

2. Test a variety of metalic items to see which are magnetic. Are any U.S. or foreign coins magnetic?

You will find a Canadian nickel to be magnetic. The U.S. nickel is not magnetic because it has a lower percentage of nickel metal.

3. Touch the point of a nail to the north pole of a magnet. Bring the point of the nail near a compass. It repels the south pole and is therefore south. You will find that an opposite pole always forms on the part of a piece of iron that is touched to a magnetic pole.

4. You can get an approximate idea of the strength of a magnet by counting the number of small nails that it can pick up as compared with other magnets.

5. Make a record of music or speech on a tape recorder. Play it back. Then move an alnico magnet near the tape. The sound is changed or may disappear when you play it back. The tape makes a record of sounds by magnetizing the brown material on the tape. When you bring a strong magnet nearby you destroy the magnetic pattern. The tape may be re-used after this.

Keep magnets away from music tapes, cassettes, video casettes, and computer disks or tapes. Do not do these experiments in the same room. You may accidentaly erase your tape.

SCIENCE PROJECT

Does the needle of a magnetic compass point exactly north?

Take a compass out-of-doors in a place far from a building and from iron objects. Mark on the ground the direction in which the dark end of the needle points when it comes to rest.

Now find the direction of true north at that spot by observing the shadow of a tall stick. Mark the position of the shadow of the top of the stick as it changes position from 11 A.M. to 1 P.M. on any sunny day. The direction of the *shortest* shadow is exactly north.

Compare the direction of north shown by the compass with that shown by the sun. Are they exactly the same? If not, how much do they differ? Is such a difference the same in different towns? In different states?

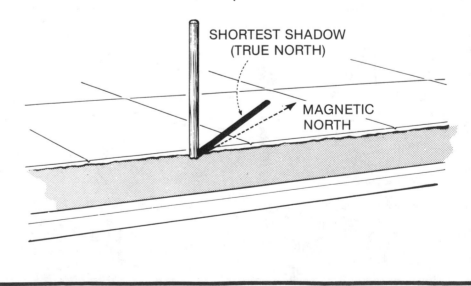

SHORTEST SHADOW (TRUE NORTH)

MAGNETIC NORTH

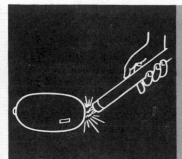
STATIC ELECTRICITY

STATIC ELECTRICITY

When lightning strikes during a storm it packs a terrific wallop. The sky lights up for miles around and thunder can be heard for 20 miles or more.

Would you like to hurl thunderbolts like these?

Well, you—can but yours will be a lot smaller. Let's find out how to do it.

Making Miniature Lightning

Rub an inflated balloon with a piece of nylon, wool or fur. Or, you can rub it on your clothing. Watch it in the dark as you bring your finger near.

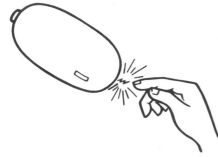

There's your miniature lightning flash! It looks like miniature lightning. It crackles like a tiny lightning bolt. It feels like miniature lightning. In fact, it is the same as real lightning, except for size. And it's made in the same way, too.

How did you make your small bolt of lightning? You rubbed a material like nylon, wool, fur, or cloth on a rubber balloon. An "electric charge" was built up on the balloon. When your finger approached, electricity in the balloon jumped to your finger and created the small lightning flash.

The word "static" means "standing still". Therefore "static electricity" refers to electric charges that remain on materials without leaking off. When a charge jumps off to cause a spark it becomes a moving type of electricity, which we refer to as "electric current".

The giant lightning flash that you see during a storm is caused by an enormous charge of static electricity that builds up as a result of the motions of billions of water droplets in a cloud. When the charge becomes big enough it suddenly jumps to the ground to create a lightning flash.

Dry Weather

Before you go around trying to put electric charges on everything you can touch, keep one very important fact in mind. The weather has a great deal to do with your ability to create charges. You will get astounding results on a winter day when it is bitter cold outside but warm and dry inside.

On a warm day in the summer, almost every static electricity experiment fizzles. An invisible film of water clings to materials and permits the charges to escape to the ground.

A Mysterious Force

Can you fill up an old nylon stocking with absolutely nothing in it? Hold the toe of the stocking against the wall with one hand and with the other hand rub it about 10 times with

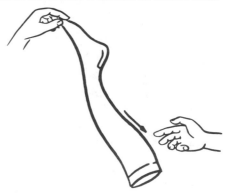

TRY THESE EXPERIMENTS

Slide across the plastic seatcover of a car on a very dry day in winter and touch the handle. What a shock you get!

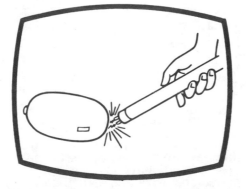

Touch the end of a flourescent tube, or small neon bulb to a charged balloon in the dark. Sparks from the balloon light up the lamps.

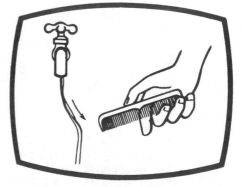

Bring a charged comb near a thin steady stream of water from a faucet. Watch the stream bend toward the comb.

STATIC ELECTRICITY

a dry, plastic polyethylene fruit or vegetable bag. You can tell that it is polyethylene if it stretches easily when you try to tear it.

Remove the stocking from the wall and let it hang freely. Watch how it blows up as though filled with an invisible leg. And notice how it clings to the wall or to your body.

On the one hand it is **attracted** by nearby objects, and on the other hand it seems to **repel** itself. Why does this happen? Let's do some experiments to find out.

Attraction and Repulsion

Blow up two balloons and tie them together with a long piece of thread. Charge each balloon by rubbing with nylon, wool, fur or clothing. Hold the middle of the string with your outstretched arm and let the balloons hang freely. Watch them repel each other.

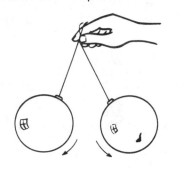

Bring your hand near one balloon. It is attracted to your hand. But push one balloon toward the other and the second balloon is mysteriously pushed away from the first.

Place your hand between the balloons. They come together and stick to your hand!

What goes on here?

Make two strips of nylon about 12 inches long and about 3 inches wide. An old nylon stocking is good for this purpose. Place both strips on blank white paper and rub them tightly with your hand. Lift up the two pieces and place them between your fingers, hanging downward. Watch how they repel each other.

Bring each charged piece of nylon near each of the two hanging charged balloons. Now the charged nylon **attracts** each of the charged balloons.

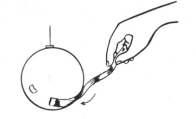

The two balloons were charged in the same way. They have charges that are alike. And the like-charged balloons repelled.

The two strips of nylon also were charged alike. And they also repelled each other. In fact, it has been found that all **like charges repel.**

But why didn't the charged nylon repel the charged balloon? **The charges must have been different.**

Plus and Minus

A charged object will either attract another charged object or repel it. Therefore there must be two different kinds of electrical charge. What shall we call them?

Benjamin Franklin, one of the founders of our nation, was the one who named the two different kinds of electricity. He gave the name **negative,** or **minus** (—) to one kind, and **positive** or **plus** (+) to the other kind.

When Franklin rubbed fur on a piece of rubber both became charged. He decided to call the charge on the rubber negative (—). He found an opposite charge on the fur and therefore called it positive (+).

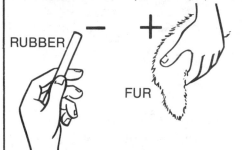

Remember this fact: the rubber balloon becomes (—) in your experiments.

When you brought the charged nylon near the (—) charged rubber balloon they attracted each other. If they had like charges they would have repelled each other. So the nylon must have an unlike, or opposite charge. It must be (+) charged.

It is easy to find out what charge is on a material. Just remember these facts:

Like charges repel.
Unlike charges attract.

Now you can understand why the nylon stocking filled up with nothing in it. All parts of the stocking were charged alike. Therefore they repelled each other and blew up like a balloon.

Cut some pieces of thin cotton and nylon thread of different lengths. Bring a charged comb near the ends of the threads and make them stand up and weave about like smakes being "charmed" by music. Notice the difference in the actions of nylon and cotton thread.

Watch from your window during a thunder strom. Observe the flashes of lightning. Why is it, a tree is dangerous to stand under during this kind of storm?

Put some small, light objects like puffed rice, cereal, insulation, sawdust or paper into a plastic cheese dish. Put the plastic cover on and rub it with nylon, wool or fur (or even with your hand). Some bits of material jump up and stick to the cover. Others jump up and down.

An Electrical World

Shuffle across a rug on a cold, dry day and touch a radiator. You get quite a shock!

Watch the spark when it is dark. Turn on the radio and listen to the "static" as you touch the aerial. You may even be able to hear the "static" when the spark jumps from your body to the radiator.

Where did the electricity come from when you shuffled across the rug? From you! You are chock full of electricity all the time. The atoms (tiny bits of material) of which you are made are composed mainly of two kinds of electricity, positive (+) and negative (—). An atom in your little finger, or in your liver, or in the tip of your nose might look like this:

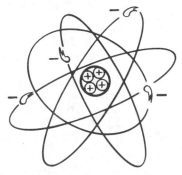

The center of the atom contains **protons** which have a positive (+)

charge. **Electrons** whirl around the center of the atom just as planets revolve about the sun, but very much faster. The electrons have a negative (—) charge. There are also **neutrons** in the center which are neutral (have no charge).

If there is so much electricity in your body all the time why don't you feel it? Scientists have found that under normal conditions there are an equal number of protons and electrons in each atom. The same is true of your body.

The (+) charge of a proton cancels out the effect of the electron's (—) charge. **Therefore, you only notice the electricity when there is more of one than the other.** In other words after you shuffle across the rug and pick up **extra** electrons you can feel the **motion** of these electrons when they jump out of your finger to the radiator. You hear a crackling sound as the spark jumps. And you see the light that is created in the spark when the electrons move.

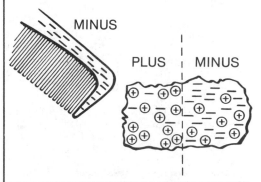

A Mystery

Charge a plastic comb by rubbing it with nylon, wool or fur. Dip it into sawdust, bits of paper or small bits of cereal. The tiny bits of material are attracted and cling to the comb.

But watch closely! One bit of material shoots off. Then another jumps off. Still another flies off as though shot from a gun.

Why do they jump off that way? If they are attracted at first, why are they later repelled?

Although the comb is charged the paper is not. It is **neutral** (has no charge). Why should a charged comb attract neutral paper?

The (—) charge on the comb repels (—) charged electrons inside the paper and pushes them to the opposite side. The paper and comb now look like this:

MINUS

PLUS   MINUS

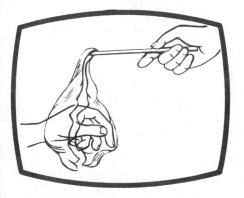

Place a long piece of plastic sandwich wrapping material. (saran wrap) against a wall and rub the surface. Then hang it over a stick. Repulsion keeps the two parts of the plastic far apart. Ask a friend to put his hand between the two parts of the plastic. The plastic wraps it up.

Bring a charged balloon over the hair on your head. Watch your hair stand up on end!

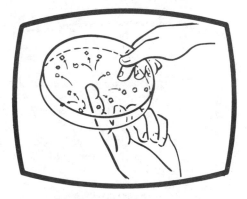

Charge up a large rigid plastic dish cover by rubbing with nylon, wool or fur. Sprinkle some bits of light material on top (sawdust, insulation, paper, ricepuffs). Move your finger around underneath. Watch the bits of material hop around or fly off.

STATIC ELECTRICITY

So the (—) charge on the comb has created a (+) charge in the paper on the side nearest the comb. The (—) charged comb attracts the nearby opposite (+) charge and the light bit of paper is then pulled against the comb.

Why do the bits of paper suddenly jump off the comb?

All electrons have like charges (—). Therefore every electron repels every other electron. When a group of extra electrons are in one place they try to push each other as far away as possible. Thus, the charge on the comb always tends to escape.

When a bit of paper touches the comb the electrons start to push each other off the comb and into the bit of paper. This happens slowly because paper does not let electrons move around easily. Finally, the paper has so much (—) charge that it is repelled off the comb by the like charge.

That's why you see the bits of paper suddenly jump off, one at a time.

Why?

Scientists always want to know WHY things happen. Very often the search for an answer does not seem to be of much practical use at the start. But later on, other scientists, inventors, and engineers find that their ability to explain why things happen also makes it possible to create new products of great value to mankind.

For example, scientists like Benjamin Franklin who experimented with static electricity two or three centuries ago could not foresee that their experiments and explorations would some day make it possible for us to produce radio and TV sets. Yet the information which they gathered, and the explanations that they developed, were later put to use by the inventors of radio and TV to build their first working models.

TRY THESE EXPERIMENTS

1. Bring a charged plastic comb near some salt on a piece of paper. Watch the salt jump up and hit the comb. Some of the bits of salt jump off almost immediately and cause faint sounds like the patter of rain. Other bits of salt stick to the comb.

How do you explain the fact that the bits of salt jump off the comb much more quickly than bits of paper?

Try other powdered materials, such as sugar, iron filings and flour.

2. Pour some salt on a sheet of paper. Sprinkle some pepper on it. Can you separate the pepper from the salt?

Bring a charged comb above the pepper and salt. The lighter bits of pepper jump up more easily and most of the salt is left behind. Repeat this a few times and most of the pepper will be removed.

3. Charge a plastic comb by rubbing it on nylon, wool, fur, or clothing. Bring it near a ping pong ball. The ball mysteriously moves toward the comb as though attracted by a magnet.

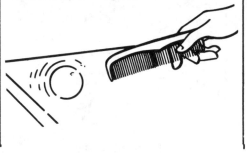

4. At night, in a dark room, pull away a piece of any kind of sticky tape from the roll. Watch for the glow at the place where the piece of tape separates from the roll. This glow is caused by static electricity.

5. Christmas decorations are often made of a very light plastic called styrofoam. Obtain a thin piece of styrofoam and rub it with nylon, fur or wool. The styrofoam can then be made to stay on the wall for a long time on a dry day.

6. Comb your hair on a very dry day. Notice how your hair won't stay down. If you look in the mirror you will see that your hair is standing on end, because it is charged with electricity.

SCIENCE PROJECT

The type of charge developed on an insulating material depends upon the nature of the material with which it is rubbed. Some materials used for rubbing will develop positive charges, others will develop negative charges.

The kind of charge can be tested by observing attractions and repulsions to a known charge. For example, a rubber balloon rubbed on clothing develops a negative charge. If the balloon is repelled by another charge, then that charge is alike and also negative. If attracted, the charge is unlike and positive.

Produce electric charges on various insulating materials, especially plastics, and find out what kinds of electric charge they produce by bringing them near a charged rubber balloon.

Does each material always produce the same charge?

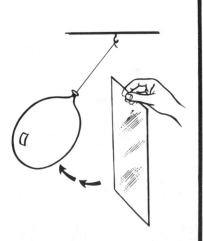

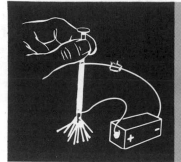

ELECTRIC CURRENT

ELECTRIC CURRENT

Can you make an alarm that will ring a bell when it rains?

You need two or three flashlight cells, a small bell (or buzzer), a salt tablet, two thin wooden sticks (such as ice cream sticks), two rubber bands and some electric wire. You can get the bell or buzzer, wire and flashlight cells at a hardware store. (A spring type clothespin may be used instead of the sticks and rubber bands.) Salt tablets are available in drug stores. See bottom of page 55 for a way to make your own salt tablets.

Make small cardboard boxes to hold the cells. Put a bolt and nut through the end of each box for making connections to the cell. (A plan for making these boxes is given on page 56.)

Mark the position of the top of the cell in each box (+), and mark the position of the bottom of each cell (—).

Use short electric wires A and B to make connections, as shown. Use long wires at C and D. (If you are not familiar with methods of connecting wires, **see page 56.**) Mount the cells and bells on a board to make a unit that can be easily carried.

Now test your rain alarm to see if the electric circuit is working. Touch the ends of the long wires together at E and F. The bell should ring.

The bell rings only if there is a **complete circuit** of electric wire from the (—) end of the cells to the (+). You **complete the circuit** and make the bell ring by touching the ends of the long wires together. You stop the ringing of the bell by separating the ends of the wires and **breaking the circuit.**

You can now make the rain alarm with a salt tablet, as follows: wind the bare ends of the long wires from

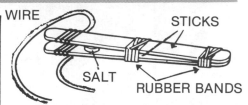

the bell and cells around the ends of two thin sticks, as shown in the drawing. Wind one rubber band around the other ends of the sticks and another around the middle. Place the salt tablet between the wires and the middle rubber band. Fasten the sticks to a long board outside the window, with the wires going to the bell inside the house. Now all you have to do is wait for rain.

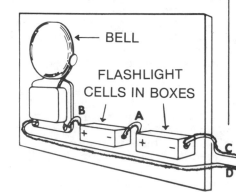

BELL

FLASHLIGHT CELLS IN BOXES

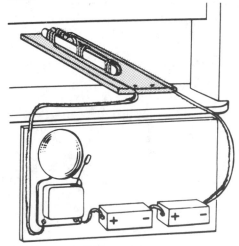

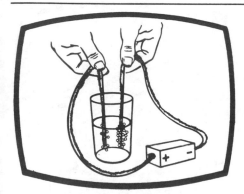

Put the ends of wires from a flashlight cell into water to which some vinegar has been added. The current breaks up the water into the two materials of which it is made, **hydrogen** and **oxygen.** You see these materials as bubbles that form on the ends of the wires.

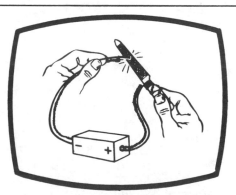

Tape the bare end of the wire from a cell to the metal handle of a nail file. Scrape a wire from the other end of the cell along the rough part of the file. Electric current heats tiny scrapings of metal to cause glowing sparks. Don't do this too often, or the cell will be ruined.

Which terminal is (+) and which is (—)? You can tell by sticking the wires from a cell into a potato, close to each other but not touching. After a minute remove the wires. The wire from the plus leaves a greenish colored hole. The other hole is not colored.

ELECTRIC CURRENT

To test your instrument before it rains, pour some water onto the salt tablet. It softens and crumbles. The rubber band pulls the sticks together. The ends of the wires touch and the circuit is completed. The bell rings. Your rain alarm works!

Wipe off the wet salt tablet, dry the sticks, put a new salt tablet between the sticks and your rain alarm is ready to work again.

Electric Pumps

You can learn a great deal about **electric circuits** by experimenting with the bell and cell boxes from your rain alarm.

Try this experiment. Turn one of the cells around and connect both (—) ends together. No matter what you do the bell doesn't ring.

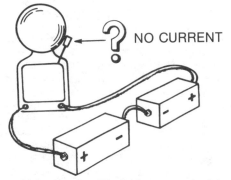

NO CURRENT

Now try it with (+) connected to (+). Again the bell doesn't ring.

But when (—) is connected to (+) the bell rings. Why?

An **electric current** is the flow of tiny, invisible electrical particles called **electrons.** The flashlight cells are like "electric pumps". They push electrons through wires to produce the electric current, which can then ring bells, light lamps, run electric motors and make radio and TV sets operate.

There are a number of ways to make the tiny electrons move in the wire. In a flashlight cell there are certain chemicals that make the electrons move. In a powerhouse, whirling magnets make the electrons move to create electric currents. In light meters, used by photographers, light makes electrons move. Scientists are now experimenting with other ways to cause electric current.

When an electric "pump" works there must be a supply of electrons to be pumped. So, electrons are taken into one end of the pump and pushed out the other. In other words every such electric pump must have an "inlet" for electrons and an "outlet", just like a water pump.

In your flashlight cell, the inlet where the electrons come in is at the top or (+) end. The outlet, where the electrons go out, is at the bottom or (—) end.

INLET

ABC CELL

OUTLET

When you connect the (—) of one cell to the (—) of the other, electrons are pumped out of both cells toward each other. So they oppose each other and no current flows.

If you connect the (+) of a cell to the (+) of another you will always find the (—) terminals connected together at the opposite ends. Again the flow of electrons from one cell opposes the other. So, no current flows.

NO CURRENT

But if you connect the (+) of one to the (—) of the other the cells help each other and current flows.

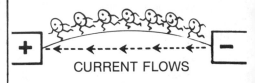

CURRENT FLOWS

Volts

Why should we use 2 cells for the rain alarm? Wouldn't 1 cell work? Try it. The ringing of the bell is much weaker, or it may not ring at all.

Now try it with 3 cells, connected together (+) to (—). The bell rings louder.

You have heard of **volts.** The volt is a unit of measure of the "push" making the electrons move. Scientists refer to this push as **electromotive force** (EMF).

A flashlight cell has an EMF of 1½ volts. When two cells are connected in **series** (one after the other) they supply 1½ plus 1½ or 3 volts. This is a greater push on the electrons and more of them are made to move. Thus the **electric current** becomes greater and the bell rings louder.

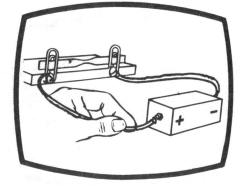

Make a **fuse** by stretching a piece of foil from a candy wrapper (or kitchen foil) between two clips on a board, and connecting the clips to a cell. A large amount of current flows, melts the foil, and breaks the circuit.

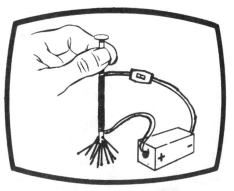

Make an electromagnet by winding many turns of thin wire around an iron nail. Connect the ends of the wire to a cell for a short while and pick up small nails with it. Always use a switch when working with electricity.

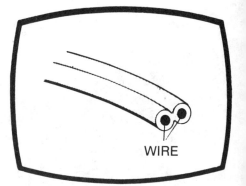

WIRE

Get a piece of extra lamp wire that is not connected to a lamp or to the house wiring. Cut it in two with scissors. Notice the two thick wires completely surrounded by insulation. Why are two wires needed?

Conductors and Insulators

Remove the salt tablet from your rain alarm. Place different kinds of materials between the ends of the wires as shown in the drawing. Try a penny, a metal washer, kitchen foil, cardboard, wood, glass, cloth, rubber.

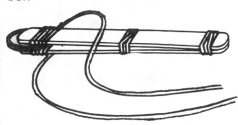

With clean metal objects between the ends of the wires the bell rings. With cloth, rubber, glass, cardboard and most other materials the bell does not ring.

Metals are good conductors of electricity and permit the electric current to flow through them, thus completing the circuit. They are said to be **conductors.** But the other materials stop the electrons and break the circuit. These materials are called **insulators.**

Pushbuttons and Switches

Use the two sticks and wires from your rain alarm, without the salt tablet. Slip a toothpick or match-

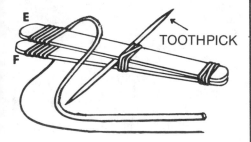

TOOTHPICK

stick between the sticks near the middle rubber band, until the two wires are slightly separated and do not touch. The bell and cells should be connected in the same way as for the rain alarm.

Now press together the ends of the sticks. The wires touch and the bell rings. Let go. The ringing stops as the bent sticks spring back and break the circuit at the ends of the wire. You have made a simple kind of **pushbutton.** Pushbuttons are used for bells because the contacts spring apart and stop the current when you let go.

Now separate the two sticks of your rain alarm by removing the rubber bands. Nail the ends into a board, as shown in the drawing. One stick (A) has two small nails or tacks so that it does not move. The other (B) has only one nail at the end, and may be rotated.

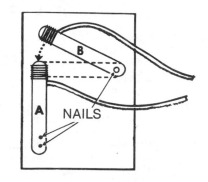

NAILS

Rotate the movable stick so that it touches the other stick, as shown by the dotted lines. If your circuit is otherwise complete the bell rings. And it keeps on ringing until you move the sticks apart, or until the cells are run down.

You have made a switch, of the kind used to turn lights on. In this kind of switch the current stays on until you move the metal contacts

away to break the circuit. You need such a switch for lights because you want the current on for a long time, without having to keep your finger on the button all the time.

Lamp Bulbs

Strip 3 inches off the end of a long piece of insulated wire. Wind a fairly stiff, long piece of bare copper wire around the metal part of the base of

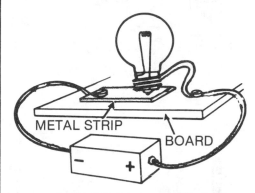

METAL STRIP   BOARD

a small flashlight lamp. Nail the other end of that wire to a piece of wood. Connect the wire to the (+) end of a flashlight cell. Connect the other end of the cell (—) to a metal strip nailed to the board. Adjust the position of the wire holding the bulb so that the tip of the bulb touches the metal strip. The bulb lights up. You have made a simple kind of socket for the bulb.

Electrons flow from the (—) end of the battery up into the metal button at the bottom of the bulb. Then they go up a wire inside the lamp, across the special wire made of the metal **tungsten.** This tungsten wire gets very hot and glows. The electrons return to the cell through another wire inside the lamp, to the wide band of metal near the bottom of the bulb, and through the wire going back to the (+) of the cell. Thus

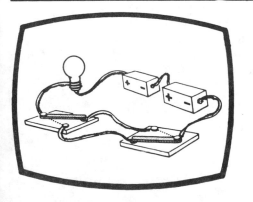

Set up this circuit to show how a light may be turned on or off from either upstairs or downstairs. Make the switches from metal strips or wires wound around sticks.

You can get a salt tablet at your pharmacy or place some table salt on a piece of wax paper. Add a few drops of water soluble school paste. Roll the salt into the paste and then let it dry. You may use this instead of a salt tablet.

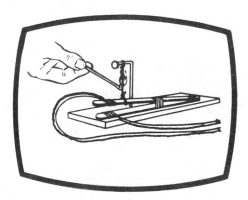

Change your rain alarm into a fire alarm by keeping the wires apart with a tight cotton thread. When the thread burns the wires are pulled together to ring the bell.

ELECTRIC CURRENT

there is a complete circuit from the cell through the lamp.

Will the bulb light if you reverse the current by reversing the connections at the cell. Try it? What do you find? See if you can explain the results of the experiment by yourself.

HOW TO MAKE BATTERY HOLDERS

Use cardboard from a cereal box or file folder. Cut out the shape shown in the drawing, measuring the dimensions as carefully as you can. Make holes with a pencil point at the center of the squares at C and D.

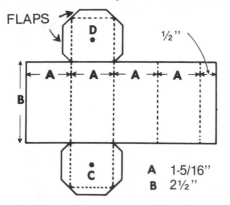

FLAPS

D

½"

A — A — A — A

B

C

A 1-5/16"
B 2½"

Cut out the figure with scissors. Crease the dotted lines with a kitchen knife, using a ruler to keep the creases straight. Fold up along the creased lines and fasten the box together with gummed tape.

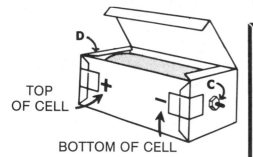

TOP OF CELL

BOTTOM OF CELL

Place a washer on a small bolt and push it through the hole at C from the inside. Place a washer on the outside and tighten the bolt with a nut. Do the same with a bolt through hole D.

Slip a size D flashlight cell into the box and close it with gummed tape. You can now make connections to the batteries by tightening wires to the bolts, using nuts on the outside.

TROUBLE SHOOTING

1. Check all connections to see that they are tight.
2. Check to see that there is no insulation or dirt under the wires where you made connections.

3. See that cells are connected (+) to (—).
4. Try new cells, if you are using old ones.
5. Sometimes an extra cell or two will be needed to make a bell ring.

HOW TO MAKE ELECTRIC CONNECTIONS

Use insulated wire, unless bare wire is called for in the directions. Cut the wire about 2 inches longer than the distance between terminals.

You must remove the insulation from about one inch of wire at each end. If the wire is covered with "enamel" (dark brown color) scrape the enamel off with a dull knife or other metal edge. Polish it with steel wool until the copper shines through.

If the wire has cloth insulation, unravel the cloth and cut it off with scissors. Sometimes you will find enamel under the cloth and this also must be removed.

Second, wind the end of the wire around the bolt only in a **clockwise** direction (in the same direction of rotation as the hands of a clock). Otherwise the wire tends to loosen as you tighten the nut.

HOOK WIRE LIKE THIS

TRY THESE EXPERIMENTS WITH AN ADULT

Connect a lamp bulb to 2 cells in series (one after the other). Why is the bulb brighter than with one cell? Then try 2 bulbs in series with 1 cell. What do you think will happen?

Examine the inside of a flashlight. Note how the cells are connected (+) to (—) when they are placed inside. Figure out how the circuit is completed, and how the switch works. Reverse one cell. Does the bulb now light?

Look at an automobile storage battery. Count the number of separate cells. Figure out the voltage of the battery from the fact that each cell gives 2 volts. How are the connections made so that the (+) of one cell touches the (—) of another?

DO NOT TOUCH BATTERY:

Break open an old flashlight cell. What is in the center of the cell? What is on the outside of the cell? Find out how a cell makes electric current.

SCIENCE PROJECT

Make a sensitive electric meter. Wind about 30 turns of insulated wire around a cardboard or plastic tube about 2 inches in diameter and about 1" to 2" in length. Leave about 12 inches of wire free at both ends of the coil. Tape the coil in place.

Make a base (a cardboard box will do) for the coil and a platform to support the magnetic compass. Mount the compass inside the coil so that it is in the center of the tube and in a level position.

Turn the base until the needle points toward the wires at the center of the windings around the tube. Your meter is now ready for use.

Connect the bare ends of wires with a flashlight cell and a tiny lamp. Tape the ends of the wire to the cell, on to the center of the top and one to the bottom. The lamp lights up. Notice that the compass needle moves and then stops at a certain point. How far does it move?

Try a different lamp bulb. Does a different amount of current flow?

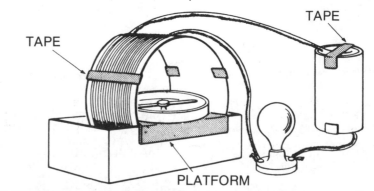

TAPE

TAPE

PLATFORM

LIGHT AND SIGHT

LIGHT AND SIGHT

Have you ever seen a solid object that can be completely invisible? You see such an object every day in the week! Yet you pay no attention to it at all!

Look through a window at night while standing outside in the dark. Look for the glass. You know it's there. You can feel it if you reach out. Yet you can't see the glass at all!

Now look at a window of a dark room from the outside. You can't see a thing inside. But the presence of the glass may now be noticed as the result of reflections from lights in the street.

You have just shown that you need light in order to see. Some materials, like glass, let light through. They are called **transparent.** In order to see something an object must "bounce" light to your eye. This bouncing of light from objects is called **reflection.**

When you stand in the dark outside the lighted room, reflection of light from the glass window to your eye is very weak, as compared with the light coming straight through. So you do not see the glass at all!

Objects in the room are visible when the lights are on because they reflect light to your eye. Your eye then captures this light and you can see the objects.

NEVER LOOK DIRECTLY AT THE SUN

Seeing Around A Corner

Put a penny into a cup that is not made of transparent glass. Move away from the cup, and down, until the edge of the cup blocks the penny from sight. Now slowly pour water into the cup without moving your head. The penny gradually comes back into view!

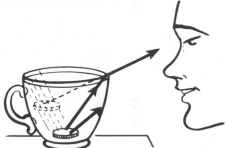

Before you add water, the light that is reflected from the penny and travels toward your eye doesn't reach you because the material of the cup stops it. Such a material is called **opaque.**

When you add water, light from the penny bends and gets around the edge of the cup to your eye. So the penny is now visible.

Now do a similar experiment with water. Hold up a glass of water and look through it at some bright object. Everything appears wavy, broken, and out of shape. Light rays from the object are bent by the glass.

All transparent materials can bend light in this way. This bending of light is called **refraction.**

TRY THESE EXPERIMENTS

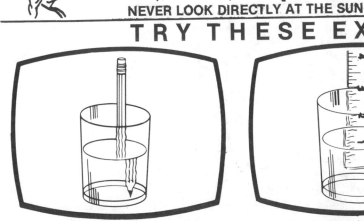

Place a pencil in water in a glass jar. In certain positions the pencil looks as though it is broken in two.

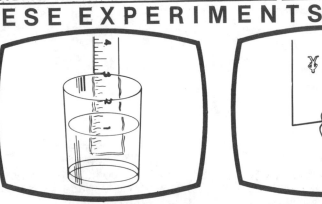

Look through a bottle of water at your ruler. The bottle acts as a magnifying glass and makes an enlarged image.

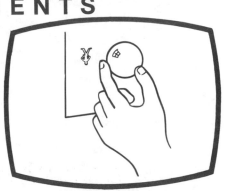

A clear glass marble acts as a lens. Hold it up very close to a wall and note the tiny image that is formed on the wall.

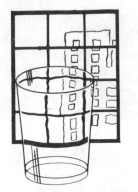

Bending Light Two Ways

Send some beams of light through a jar of water and see the way in which they bend. Set up a flashlight, a card with several vertical slits, and a jar of water on a piece of white paper, like this.

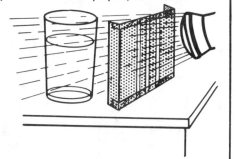

An easy way to make the slits is to cut several 4 inch lengths of card, about ¾ inch wide, and then use gummed tape to fasten the tops and bottoms together, with a slight separation between strips. A bend in the card at each side enables you to stand it on edge. Point the flashlight slightly downward so that the beams of light are seen on the paper. Look down from above. This is what you see.

Add a small amount of soap to the water. The increased reflection of light from the soap particles makes the beam more visible.

Notice how the water in the jar bends the rays of light from the flashlight so as to **focus** them to a point.

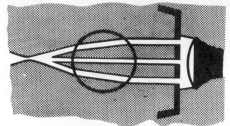

Now try this experiment. On a sunny day, outdoors, move a magnifying glass toward and away from a sheet of paper placed in a sunny spot. At a certain position the rays of sunlight focus to a point.

Let the rays focus onto a dark spot, such as an inkspot. The black color of the inkspot absorbs the rays of the sun. The paper soon begins to smoke and may even catch fire. If this should happen be ready to put it out quickly.

CAUTION: Do not place your hand between the magnifier and the paper.

Stir up some dust in the path of the light passing through the lens. Dust particles reflect some of the sunlight to your eye and you will be able to see the beam.

The lens and the jar of water focus rays of light in a similar way. But the lens does a much better job, because it is designed for that purpose.

Forming Images

Hold up a magnifying glass near a wall opposite a window and move it back and forth. When the lens is at a certain distance from the wall you see an **image** of the window appear on the wall. If your friend walks past the window you see a small picture of him "walking" upside down on the wall! In fact, you can see images of houses, trees, cars, and other objects, all upside down (inverted).

Here's why you can see the image of your friend—even though it is inverted on the wall. Suppose that your friend is standing in sunlight in front of your lens. Light from the sun strikes his head and bounces off. Some of it reaches your lens and passes through. The lens bends the rays and brings them to a focus on the wall, like this.

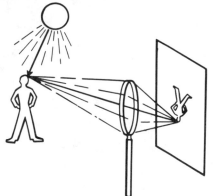

Move your head to one side as you look out of a window. Imperfections in the glass show up as a wavy appearance of distant objects. This effect is caused by refraction.

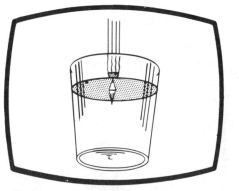

Put the point of a pencil just beneath the surface of water in a glass and view it from below. You see a mirror image of the point, while the pencil above the water is not seen at all.

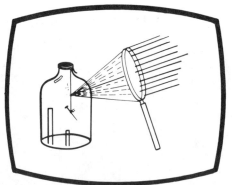

Use a lens to focus rays of sunlight onto a black thread holding a nail in a bottle. The thread burns and the nail drops. But it won't work if you use a white thread.

Some of the rays of sunlight strike his feet and are reflected toward the lens. They pass through the lens and are brought to a focus like this.

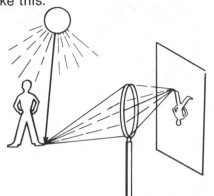

Notice how rays from his head focus on the lower part of the wall, while rays from his feet focus on the upper part of the wall. As a result an inverted image is formed.

Lenses are used in cameras because they form images. The lens is at the front of a camera and produces an image on the film at the back. When developed, chemical on the film blackens wherever light has struck, and thus captures the image to make a picture.

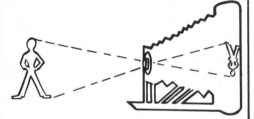

How You See

Walk into any room and look around. It is certain that there are a couple of lenses around. Where? Right in your eyes!

Each one of your eyes is like a small camera. A transparent, bulg-

ing part near the front of the eyeball serves as a lens. It forms an image on the **retina** at the back of the eyeball.

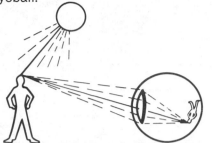

When light reaches the thousands upon thousands of nerves on the retina, electric currents race up the nerves to your brain. These electrical messages form a pattern that the brain "reads" to tell you what kind of image is on the retina, and therefore, what kind of object you are looking at.

But just a moment. The image in your eye is upside down! Then why do you see things right side up?

Your brain learns by experience what the different electric current patterns from your retina mean. When you were a few days old you saw nothing but light and dark, without any meaning at all. You gradually learned to connect certain patterns of light and dark with certain objects in their right positions. After a while you could "see" the objects and recognize what they were.

Magnifying Images

As a result of the studies which scientists have made about light they have found many ways to improve your ability to see. For example, a magnifying glass helps you to see tiny objects better. It is so simple that you can actually make one out of water!

Simply put a piece of waxpaper on newspaper and place a drop of

water on the waxpaper. Then newspaper print appears greatly magnified. Try this with a drop of glycerine. It works still better.

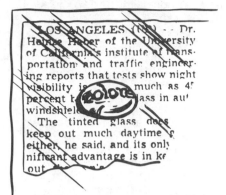

Why does the drop magnify? It has a bulging shape. Such a shape is called **convex**. A magnifying glass also has this bulging shape. Suppose that you are looking at the letter "t" on a newspaper through a bulging drop of water or a lens. Light rays from the top and bottom of the "t" are bent by such a convex lens in the following way.

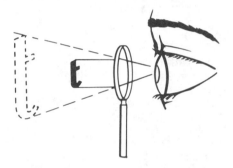

As a result you think that the light from the top of the "t" came from a point higher than it really did. At the same time you think that light from the bottom of the "t" came from a point lower than it really did. So the image that you see widens out and looks much bigger than it really is.

When you look at a fish tank from one corner you can see one fish appear as two. Both sides of the tank create different bending angles to cause a double image.

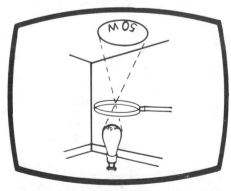

Bring a hand magnifying glass above the bulb of a table lamp. The lens acts as a projector and forms a greatly enlarged image of the markings of the bulb on the ceiling.

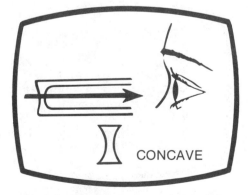

CONCAVE

The bottoms of thick drinking glasses are often **concave** (curve inward). Look through the bottom of such a glass and note that everything appears right side up and smaller.

Why Does Light Bend?

Like a good scientist you should have been wondering **why** light bends when it passes through a transparent material.

Scientists have discovered that in some ways light acts as a wave. The speed of this wave in air is about 186,000 miles a second, fast enough to go around the earth in the time it takes to wink your eye!

But light waves slow down as they travel from air into materials such as water and glass. If one part of the wave strikes the surface first it slows down, while the other part is still traveling at its normal speed in the air. So the whole wave swings around and changes direction.

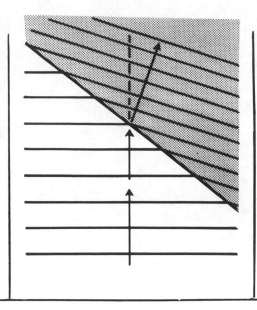

You might compare this moving wave with a line of marching soldiers in a parade. If the marchers on one side of the line slow down, the other side of the line swings around. The whole line then changes its direction. In fact, that is exactly how a line in a parade goes around a corner.

Without this simple slowing down of a light wave in a transparent material, lenses would not form images, and we could not see as well as we do.

* * * * * * * * * *

Our knowledge of light has enabled us to make eyeglasses, telescopes, microscopes, projectors, and cameras. These instruments help make your life healthier and happier, and have added much to our knowledge of the world.

TRY THESE EXPERIMENTS

1. Make a wave in a large flat pan with water in it. Push the water with a ruler or wide piece of cardboard. A wave travels outward and reflects off the side of the pan. A light wave is thought to reflect in a similar manner.

2. While riding in a car on a sunny day watch for "images". Pools of water appear in the road ahead and disappear as you approach them. This illusion is caused by bending of light rays by warm air near the ground. It looks like water because it seems to give a bright reflection of

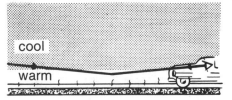

the sky. Look for upside down images of cars in these false pools of water.

3. Look through the lenses of a pair of eyeglasses held at arm's length. If the wearer of the glasses is farsighted you will probably see upside-down images in his **convex** lenses. If he is nearsighted you will see small upright images in his **concave** lenses.

4. Watch the streamers of light that form on the bottom of a shallow stream or pond while the sun is shining overhead. They are caused by the curved surface of the wavy water on top of the stream, which refracts the light as a lens does.

5. The shadow of a "water bug" that walks on the surface of water appears on the bottom of a shallow stream with large dark ovals around

each leg. These dark areas are caused by the fact that the water is depressed under each leg by the weight of the bug and forms a concave "lens" of water. This concave shape then bends water away from the area and causes the large dark oval.

6. Watch the sun as it goes down below the horizon at sunset. Notice how it appears flattened. This is caused by bending of light from the sun as it passes through the air from outer space. The day is actually lengthened a few minutes each day because of this refraction of light.

7. Make a simple telescope by placing a concave lens near your eye and a convex lens beyond it. Move the convex lens toward and away from the concave lens until you see a magnified image.

SCIENCE PROJECT

You may be able to project an enlarged image of a person on a wall by shining a strong light at his face and placing a lens nearby. Move the lens back and forth until the image is in focus on a wall.

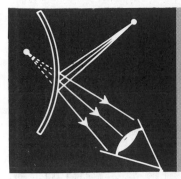

MIRRORS

MIRRORS

Would you like to multiply your money?

You can get the illusion of becoming richer by using two pocket mirrors. For this interesting trick it is best to have mirrors without frames. You can often buy such inexpensive mirrors in local stores. If available only with frames, carefully remove them using screwdriver and pliers. Always tape the edges of mirror with transparent tape to protect your fingers and face.

Use gummed tape on the backs of the mirrors to hinge them together. Then stand them up as shown in the drawing. Place a quarter on the table between the mirrors. Then, as

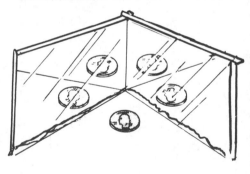

you slowly bring the free ends of the mirrors closer together, additional quarters come into view. Your money seems to grow to $1.00, then to $1.25, to $1.50, to $1.75, to $2.00. Your increase in money is finally stopped when both mirrors touch the quarter.

Why does this happen? Let's find out by studying the way in which mirrors form images.

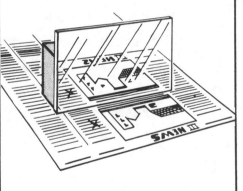

Use gummed tape or a rubber band to attach one mirror to a block of wood or small box, so as to make the mirror stand on edge on the table. Place the mirror on a piece of

newspaper with the edge along one of the lines of print. Make a pencil mark on one of the lines of print in front of the mirror.

Note a number of facts. First, an **image** of the printed lines appears in the mirror. Second, the lines in the image seem to be the same size and shape as the lines on the newspaper. Third, the printed letters appear reversed in the image. Fourth, the image of the pencil mark appears to be as far back of the edge of the mirror as the pencil mark is in front of the mirror. Except for reversal, the image seems to be a perfect duplicate of whatever is in front of the mirror.

What Causes the Image?

Cut two strips of thin cardboard 2 inches long and ¼" wide. Cut two squares 2" on each edge. Use gum-

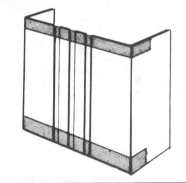

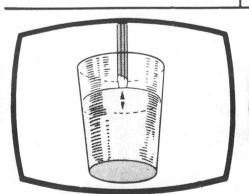

Look at the tip of a pencil that is pushed under the surface of water in a glass. View it from below and to one side. An excellent mirror reflection of the pencil point is seen in the upper water surface. Look for similar reflections in a rectangular fish tank.

Look into a store window on the shady side of a street on a bright day. The window acts as a mirror and shows reflections of people and objects in the street. Are reflections seen at night?

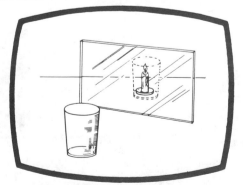

Place a pane of glass vertically on a table. Put a glass tumbler in front of the pane of glass, and a lit candle an equal distance behind the pane of glass. When viewed in a dark room the candle appears inside the glass tumbler.

med tape to fasten them together to make a group of narrow slits, as shown in the drawing. The squares are bent to make a stand for the slits.

Shine the light from a small flashlight at the slits so that several beams of light are created. Let the beams strike a mirror. You see that they are **reflected** (bounce off).

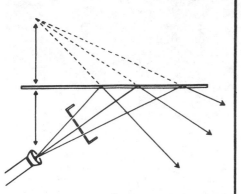

Notice how each beam reflects from the mirror at the same angle at which it strikes.

Look into the mirror. See how the beams seem to be coming straight from a point back of the mirror. This is the image of the lamp in the search-light.

Try rolling a ball along the floor so as to hit the wall from different angles. The rolling ball acts like a beam of light and bounces off at the same angle at which it strikes.

We see an object because of the pattern of light **rays** (narrow beams) that come from it and enter our eyes. When these light rays are reflected from a flat mirror all the angles remain the same afterwards, as before. So, the pattern remains the same, and the image appears to have the same size and shape as the original.

But the directions of these rays of light are changed. As a result we see them coming from a new place, behind the mirror.

Why is the image reversed? The drawing shows 3 narrow rays of light striking a mirror. They are in the order 1, 2, 3. Notice that after reflec-

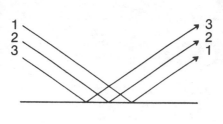

tion the order is reversed to 3, 2, 1. Thus, a reflection reverses the order of the rays. We then see the image in reverse position.

Mirror Surfaces

Why are certain surfaces shiny and mirror-like?

Put some water into a flat dish. Notice how it gives mirror images of your face and objects in the room.

Put your finger into the water and make some waves. Now the surface becomes disturbed, and is no longer flat. The image jumps about and becomes unrecognizable.

Let the water become still once again. The mirror image is seen as the water becomes smooth.

Mirror images are observed only when a surface is smooth. In that case the rays are reflected in a regular manner and keep the same pattern that they had before striking the mirror, as shown in the drawing.

On the other hand, if the surface is rough the rays are reflected in many different directions and the pattern of rays is broken up.

This type of **diffuse** reflection is important in getting proper illumination in a room. Mirror-like reflections from smooth surfaces are very disturbing to the eye. Walls and ceilings are therefore usually covered with surfaces that are not too smooth and therefore give diffused light.

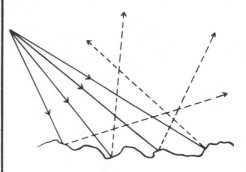

As a result of the irregular reflection from rough surfaces and scattering of light rays in all directions, it is impossible for mirror images to form. Surfaces that are just a bit roughened will show unclear mirror images, that appear as "glare" or shiny reflections.

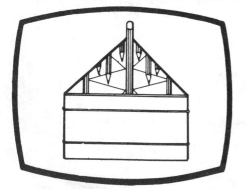

Make a hall of mirrors. Use rubber bands to hold three equal-sized mirrors together. An object placed inside the three mirrors appears to be multiplied many times over inside a great hall of mirrors.

Look for curved mirror images in the chrome parts of a car. You will also see similar images in the body of a car if it is highly polished.

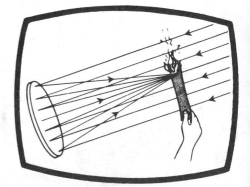

Use a concave shaving mirror to focus the sun's rays on a piece of dark paper. If held at the focus (point where the rays meet) the paper may be set on fire. Do this outdoors to avoid danger of fire.

Many Images

Set up two mirrors facing each other and about 2 inches apart, as shown in the drawing. Place a pencil between them. A parade of pencil images appears in the mirror behind the pencil.

Rays of light from the pencil can reach your eye in many different ways. One group of rays can reach your eye directly. Another can reach

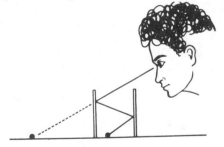

the eye after bouncing off the back mirror. Another can come by way of a reflection from the front mirror followed by a reflection from the back mirror. Other paths with 3, 4, 5 and more reflections are possible. Each of these different paths produces a different image of the pencil. Therefore, we see the parade of pencil images.

Turn the front mirror a bit. Now you see a curved parade of pencils. Change the angle a bit more. The parade curves more sharply.

You can see why the first experiment with two mirrors at an angle produced a number of images of a quarter. In addition to a group of rays that arrives from the quarter directly, there are a number of different ways in which rays could come by reflection. One image is formed by rays that are reflected once from one mirror. Another is formed by reflection from the other mirror. Still another is formed by a double reflection from both mirrors. If the angle between mirrors is large only a few such reflections are possible. But if the angle is small, as happens when you bring the mirrors closer together, a greater number of reflections are possible. As a result, more images are seen.

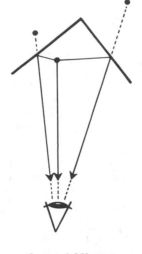

Curved Mirrors

All of your experiments up to this point have been performed with flat (or plane) mirrors. What happens if the mirrors are curved?

A simple way to study curved mirrors is to use a shiny tablespoon. Hold the **convex** (outward bulging)

part of the spoon toward you. You see a tiny image of yourself in the spoon. If you hold the long side of the spoon vertically, a long, thin image is observed. If the spoon is held horizontally your image appears short and fat.

The drawing shows the way in which a curved surface changes the pattern of rays reflecting from a shiny spoon, as compared with a flat mirror. This changed pattern causes the image to seem much closer to the mirror. The image is also jammed into a narrower space and therefore seems smaller. Distortions in the image are caused by any distortions in the shape of the mirror.

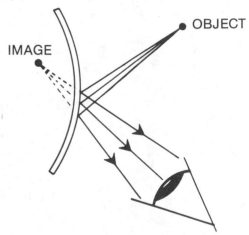

IMAGE   OBJECT

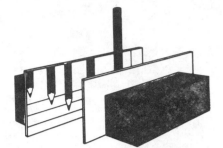

Try writing your name or tracing a simple figure while looking into a mirror. Use a book or cardboard to hide your writing hand from direct view. It is very difficult to do this kind of "mirror writing".

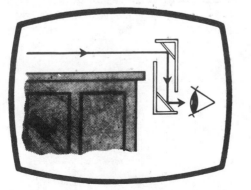

Make a periscope. Use two mirrors in a long cardboard box or tube, angled as shown in the drawing. Use it to see objects around corners.

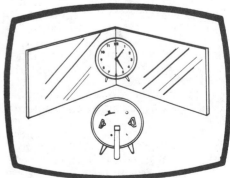

View the face of a clock in two mirrors at right angles to each other on the table. Unlike single mirrors, which give reversed images, this arrangement produces images in correct position. Look at your face in this mirror.

The ability of convex mirrors to produce small images of large objects makes them very useful as rear view mirrors in buses and automobiles. The driver can then see a large area behind him in a compact mirror.

Now try this experiment with the shiny spoon. Turn it around so that the **concave** (inward-curving) side faces you. This time you see a small

upside-down image of your face. The drawing shows why this happens. A concave surface causes the rays coming from an object to meet and cross after reflection. This turns the rays upside-down to cause an upside-down image.

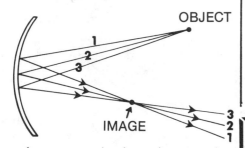

OBJECT

IMAGE

A concave shaving mirror can be used to show other interesting facts. Hold such a mirror close to your face. You see an enlarged, erect (right side up) image of your face. It is not an upside-down image because the rays have not yet met and crossed.

Slowly move away from the mirror while watching the image of your face. The image gets larger and larger and finally seems to fill up the entire mirror. You have reached the place where the rays are meeting.

Keep moving away. A short distance later your image reappears. But it is now upside-down and quite small.

If you study the image carefully you will find that it seems to be closer to you than the mirror. It is located at the point where the rays meet and cross.

This meeting point is very important. It is called the **focus**. A **real image** is formed at this point, one that you can actually capture on a screen. Try it. Place the shaving mirror on a table across the room from a table lamp, with the mirror facing

the lamp. Stand behind the mirror and look at a narrow strip of white paper that you move back and forth near the mirror, and between the mirror and the table. At a certain spot a clear upside-down image of the lamp appears on the paper.

This real image is a different type than the ones formed by flat and convex mirrors. No matter what you do it is impossible to form an image on a sheet of paper with a flat or convex mirror without using additional optical equipment. With flat and convex mirrors, and for some positions of the concave mirror, you see the image only by looking into the mirror. The image appears to be behind the mirror. But when you actually go behind the mirror to try to capture the image on a screen you can't find it. Images of this type are known as **virtual images.**

The special abiltiy of a concave mirror to form real images makes it possible to use them for telescopes. The giant astronomical telescope at Mount Palomar has a concave mirror that is 17 feet in diameter, wider than most of the rooms in you home.

Try This Experiment

A simple **kaleidoscope** may be made by hinging two narrow mirrors along the long edge with gummed tape. Look between the mirrors at an object held near the opposite end. Your finger, a colored page, and almost any object show beautiful patterns. These patterns are easily changed by adjusting the angle between mirrors.

TAPE ALL EDGES FOR SAFETY

SCIENCE PROJECT

If you hold a mirror in a beam of light the beam can be reflected into any nearby spot by tilting the mirror properly. If you use a combination of three mirrors formed into a corner shape, then the beam is reflected in a very special way—it returns to the source of the beam in the same direction from which it came. In other words, the corner-shaped mirrors reflect the beam back toward itself.

Show that this is so by making a corner mirror and tracing the path of a beam entering the combination of mirrors.

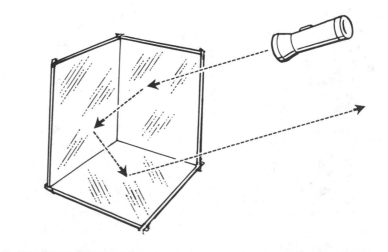

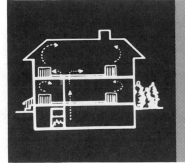

HEAT

HEAT

Can you make water shrink?

It's easy. Place a soda bottle in the sink and let hot water run into it up to the top. Turn off the water when the bottle if full.

Watch the top surface of the water carefully. You will soon see the level of the water drop. After 10 minutes it will be down about a half inch or more. The water has shrunk.

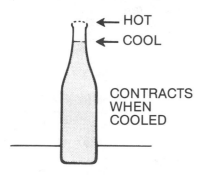

← HOT
← COOL

CONTRACTS
WHEN
COOLED

Expansion by Heat

Can you make the water return to its original size? Put the bottle of water into a large pot that contains some water. Heat the pot gently, on the stove. The water inside the soda bottle soon begins to rise. It may even reach the top and spill over.

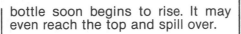

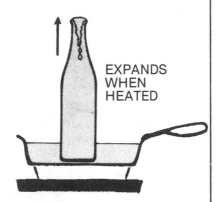

EXPANDS
WHEN
HEATED

With very few exceptions, materials **expand** (become larger) when heated, and **contract** (become smaller) when cooled. So, when the hot water in your soda bottle cooled off it took up less room and shrunk. When you heated it in the pot the water expanded and returned to its original size.

Other liquids, like gasoline and kerosene also expand when heated. Solid materials like cement roads, steel bridges and copper wires expand when heated. Air and other gases do the same.

Automobiles move because of expansion. Gasoline burns inside the engine. The gases that form are expanded by the heat. The force of this expansion makes the engine turn. The engine then makes the back wheels turn to move the car.

Temperature

Place three pots in the sink. Fill the pot on the left with warm water. Put ice cubes into the pot on the right and add water. Mix cold and warm water in the pot in the center, until the water feels neither cold nor warm.

Place one hand in the warm water and the other in the cold water. After a minute put both into the center pot. Now the hand that was in warm water feels cool. The hand that was in cold water feels warm. Yet both hands are in the same pot! Each hand feels a different temperature.

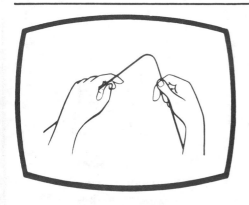

Bend a piece of wire rapidly back and forth until it breaks. The broken part gets very hot because its molecules are made to move more rapidly by your motion.

Place your hand above and below a pot containing ice cubes. Your hand feels colder under the pot than on top because cold, heavy air falls. Handle with gloves.

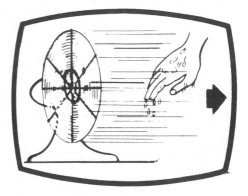

Wet your hand and place it in front of the air stream from an electric fan at least 12 inches from fan. Your hand feels cool because of the evaporation of water. Refrigerators make use of this principle.

HEAT

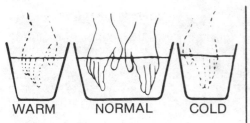

WARM   NORMAL   COLD

You are not a very accurate judge of temperature. It is better to use a **thermometer** (temperature meter).

Touch the bulb of a thermometer with your finger. The liquid in the narrow tube expands and rises. Touch the bulb to an ice cube. The

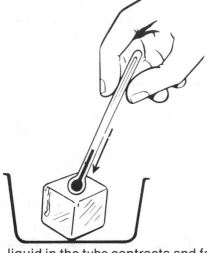

liquid in the tube contracts and falls. The thermometer measures temperature by the amount of expansion of its liquid when heated, and by the amount of contraction when cooled.

Moving Heat

In how many different ways can you cook a frankfurter?

One way is to cook it in a frying pan, with a bit of oil. Place the pan on the stove and heat it gently. The bottom of the pot is heated by the flame. The heat then comes through the solid metal of the pan by **conduction** and cooks the frankfurter.

CONDUCTION

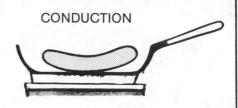

Now try this method. Spear the frankfurter on a long stick or fork and hold it above a fire. Keep it out of the direct flame. The frankfurter now gets cooked mainly by the upward flow of hot gases. This kind of flow caused by heat is called a **convection current.**

CONVECTION

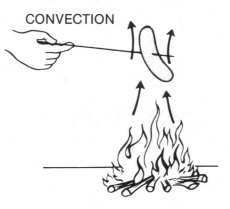

Convection currents are caused by expansion. When air is heated it expands and becomes lighter. Cooler air around it then moves in and pushes up the lighter air. In turn, this cool air is warmed and pushed up. A steady flow thus continues.

Winds are caused in this way. A sea breeze is simply the air from the cool ocean moving in toward land to push up the lighter warm air over the hot ground. The **Trade Winds** near the equator are caused mainly by cool air moving toward the equator to push up the hot, light air.

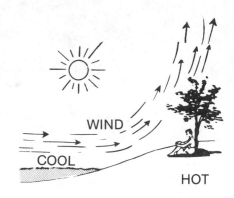

WIND

COOL   HOT

Convection currents occur in all liquids and gases. For example, the Gulf Stream, which warms up Europe, is a giant convection current of water in the ocean.

Radiation

Now try cooking your frankfurter by a third method. Place it underneath the glowing hot wires of an electric broiler. The frankfurter

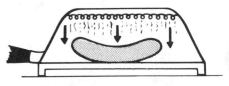

RADIATION

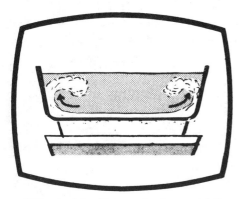

Pour cold milk gently down the inside of a pot of warm water on the stove. If the water is quiet the cold heavy milk will settle to the bottom. Turn on the heat. Watch the warmed milk rise along the sides.

DO THIS WITH AN ADULT
Set fire to the black print of a newspaper by focusing the infrared rays from the sun with a lens. The white part of the paper does not burn as easily because it reflects the rays. Do this experiment outdoors.

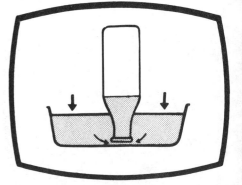

Fill a bottle with hot water. Turn it upside down over a pot. After the water spills out lower the neck of the bottle into the water. Watch the water rise as the warmed air in the bottle cools and contracts.

can't get cooked by convection because the hot air near the wires rises. Conduction of heat downward through the air is very slow because air is a **poor conductor** of heat.

In this case the frankfurter gets cooked by **radiation.** Invisible **infrared** rays are given off by the glowing hot wires of the broiler. When these rays hit the frankfurter they cause enough heat to cook it.

Heat comes to us from the sun in a similar way. Infrared rays from the sun hit the earth, warm it up, and thus make it possible for us to live.

The three methods of moving heat from one place to another are put to work in your home. Hot air, hot water or steam rises by convection from the furnace in the basement and brings heat to the rooms upstairs.

If there is a radiator in your room then the heat comes through the solid metal by conduction. The radiator then radiates infrared rays to warm up the room.

Some of the heat is sent around the room by convection, as the warm air near the radiator is warmed and rises. Distant parts of the room are then warmed up by this circulation.

Heat Energy

Can you make a candle out of butter? Melt some butter in a small jar cover, placed inside a pot containing a small amount of water. Heat

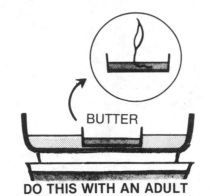

BUTTER

DO THIS WITH AN ADULT

the pot gently, as shown in the drawing. Remove the jar cover when the butter melts. Dip a short piece of string into the melted butter. When the butter hardens it locks the string in place.

Light the candle by heating the string with a match flame. Your butter candle starts to burn just like the candle you buy in a store.

This experiment shows that butter has **energy.** You could boil water with the heat from this butter candle and make the steam go into a steam engine to do work.

Why do you eat butter? Your body burns up the butter you eat and uses its energy to keep warm and to make you move.

The energy of heat is measured in **calories.** Scientists measure the number of calories in a food by drying it and then burning it. They use this heat to warm water. They can calculate the number of calories from the amount of water and the rise in temperature.

What is Heat?

Pour a small amount of perfume into a clean ash tray. Stand about 5 feet away and smell the air. You soon get a whiff of the perfume. Move further away. A short time later you smell the perfume at the greater distance.

Look at the ash tray after a few minutes. It is dry. The liquid has

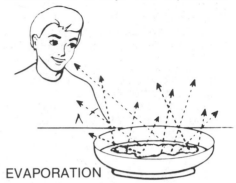

EVAPORATION

evaporated. Tiny bits of the liquid, called **molecules,** jump out into the air. They move in all directions. Some reach your nose and you smell them.

Every material is composed of tiny bits, or molecules. A molecule of water is the tiniest bit of water there is. A molecule of iron is the tiniest bit of iron. They are so small that about one hundred million of them would make up a line only one inch in length.

Molecules are always in motion. As a material is heated its molecules move faster. In other words, heat is the motion of molecules.

Why do materials expand when heated? When an object is heated its molecules speed up, bang into each other harder, and push each other further apart. So it expands.

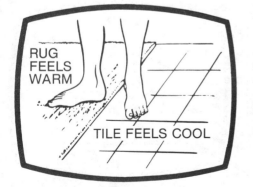

Stand barefoot, with one foot on a rug and the other on tile (or wood). The tile feels colder, even though it is at the same temperature, because it conducts heat away from your body more rapidly.

DO THIS WITH AN ADULT

Hang a paper lunch bag filled with water over a burning candle. Explain why the bag does not burn. What happens to the water?

Place the center of a rubber band against your lips and pull both ends apart quickly. It feels warm because your motion causes the molecules to move faster. Relax the rubber band. Now it feels cool.

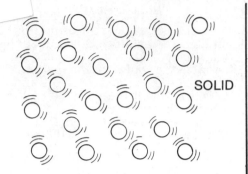

SOLID

When the object cools off its molecules slow down and are able to pull closer together. So the object contracts when cooled.

Why is heat conducted right through solid metal? The molecules of the metal on the bottom of a pot are made to move faster by the hot air that rises from the flame. These fast-moving molecules hit the ones above them and make them move faster, too. So the fast motion of molecules, or heat, is passed right through the solid metal from one molecule to the next. Finally the upper side of the metal becomes hot.

Why does water boil when heated? Heat makes the molecules of water move faster. They move so fast that they jump right out into the air. Thus the water boils away.

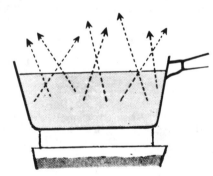

Why does water freeze when cooled?

Molecules tend to attract each other and pull together. You can see this attraction by dipping a pencil in water. When the pencil is taken out of the water it is wet. The water molecules are attracted by the wood molecules of the pencil.

When you cool water the molecules slow down. Finally, they slow down so much that the pulls of nearby molecules are enough to keep them in one place. Now the molecules of water are frozen into one position. The water has become ice.

All solids become liquids or gases when heated because the faster motion of molecules makes them break away.

The Importance of Heat

If it is too hot or too cold living things die. We are lucky that the

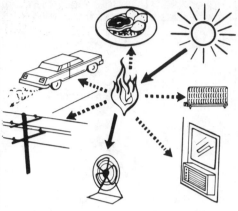

earth gets just the right amount of heat from the sun so that we can live fairly comfortably.

In winter we burn fuels to keep us warm. In summer we use air conditioners and electric fans to take heat away and keep us cool.

We eat food each day partly to supply us with the heat energy needed to live.

We burn coal, oil and gasoline to run our automobiles, trains, ships, buses and trucks. Such fuels also operate many of the powerhouses that make electricity.

You can see that knowledge of heat is very important in understanding our world and in making it a better place in which to live.

TRY THESE EXPERIMENTS

1. Add a cup of very hot water to a cup of ice in a pot and mix thoroughly. Does the hot water have enough heat to melt all the ice? This experiment will give you some idea as to the cooling ability of ice.

2. Gently pour a small amount of ink into a jar of quiet water. Watch how the color slowly moves to all parts of the liquid as the molecules move around. How long does it take to color the water completely?

3. Remove the head of a match. Hold the wooden part of the match in one hand and a nail in the other. Touch both to an ice cube. The nail soon feels much colder because it conducts heat out of your hand more rapidly.

4. Push the open end of a balloon over the open end of a soda bottle. Let hot water pour over the bottle in the sink. Expansion of air causes the balloon to blow up a bit.

5. Place your hand above and below warm water in a shallow pot. Your hand feels warmer above the pot than below because warm air is pushed up by cooler air around the pot.

6. Hammer a piece of thick wire or a nail against a heavy iron object, such as an anvil. Feel the heat in the wire caused by molecular motion.

7. Make a paper pinwheel and suspend it above a hot radiator using a string attached to its center. Rising warm air makes the pinwheel rotate.

SCIENCE PROJECT

Did you know that it is possible to make electric current flow just by heating metal wires? The method is simple.

Get two different kinds of wires, such as iron and copper. Twist the ends with pliers until they make close contact. Connect the other ends to a sensitive meter. A type known as a galvanometer will work well.

Use a match to heat the wires where they meet. Does the meter show electric current flowing?

The joined part of the two wires is called a *junction*. The arrangement of the junction to give electric current with heat is called a *thermocouple*.

Read about the thermocouple in the library.

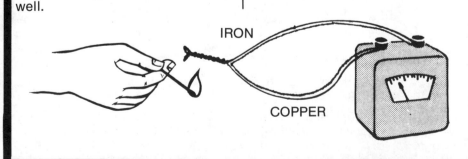

IRON

COPPER

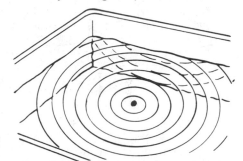

SOUND

SOUND

How would you like to make music with nothing but a string and a box?

Here's how you can do it. Get an open box with fairly stiff sides. A milk container with one side cut out works nicely. Punch two holes in one end of the container and tie a string to it, with a button on the other end. Tighten the string with one hand, as shown in the drawing, while you pluck the string with the other. Surprise! You hear music.

Play different tones by making the string tighter or looser. Try different kinds of string and thread. Nylon thread or fishline works so well that you can actually play a tune after a bit of practice.

What causes the musical sound?

Making Sounds

Make a noise by rapping the table with your knuckles (not too hard).

Blow a stream of air with your mouth and place your finger in front of the stream. You hear a noise. Rub your foot against a hard floor to make another kind of noise. Tap a thin glass with a spoon. Now you hear a tinkling musical sound.

In every case you make a sound by **doing something.** Sound is always connected with **motion** of objects or materials.

When two objects or materials strike each other they begin to **vibrate** (move back and forth) rapidly, faster than the eye can see. This vibration shakes the air and sets it into motion. The vibration of the air moves outward in the form of a **wave.** This wave reaches your ears and vibrates a flexible skin in your ear, called the **eardrum.** The **inner ear** and your brain then respond and hear the sound.

SOUND WAVE

Making Waves

In some ways sound waves resemble water waves. You can learn something about sound waves by watching water waves.

Let water run into a stoppered bathtub or sink until there are several inches of water. Make water waves by letting drops of water fall

from a wet pencil or by dipping your finger into the surface. Notice how a circular wave moves outward from the place where the drop of finger touched the water. Several waves are seen, one following the other, outward from the center. The falling drop causes a vibrating motion of the water which we see as a wave.

A picture of the water waves from the side would look like this.

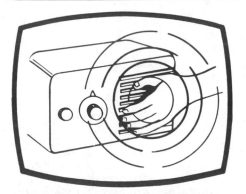

Turn a radio on loud. Place the palm of your hand in front of the speaker. Feel the vibrations of the air with your hand.

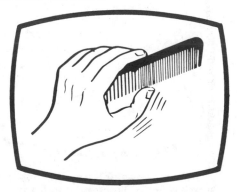

Rub the teeth of a comb with your fingers or against a hard object. You can imitate the sound of a cricket, a duck or a barking dog.

Make a stethoscope from a large funnel and rubber tube. Use it to listen to your heartbeats. Try listening to the sound of an automobile engine.

SOUND

If you watch the water waves carefully you will see some of them **reflect** (bounce) off the sides of the tub. You will also notice that two waves can move through each other in opposite directions without stopping each other.

Different Sounds

The vibrating string on your homemade milk carton instrument gave a musical tone. But when you rapped the table with your knuckles you heard a noise. Why? And why did the tones change when you tightened or loosened the string?

Place a thin wooden ruler or metal hacksaw blade on the table with most of it sticking out past the edge. Hold it firmly with one hand while

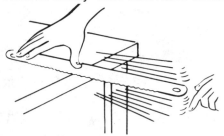

you pluck the free end with your finger. You hear a noise. Shorten the piece sticking out over the table edge and try it again. Continue to shorten it. Soon you begin to hear a low musical sound. As you shorten it the **pitch** of the tone becomes higher and higher. Why does this happen?

A musical sound is a **regular** vibration while a noise is **irregular.** If an object is springy and free to move back and forth it can vibrate in a regular manner. When you pluck the ruler or hacksaw blade only the piece sticking out beyond the table

edge can vibrate freely. That is the part that makes the musical tone.

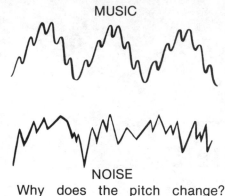

MUSIC

NOISE

Why does the pitch change? Short objects vibrate more rapidly than long ones. Therefore the shorter pieces sticking out over the table edge vibrate faster. This causes more sound waves to be sent out into the air every second. The **frequency** is then higher. And our ears hear a higher frequency wave as a musical sound of higher pitch.

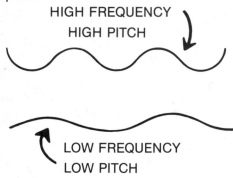

HIGH FREQUENCY
HIGH PITCH

LOW FREQUENCY
LOW PITCH

Why did you hear no musical tone at all when a long section of ruler was vibrated? The lowest tones that your ears can hear are about 16 vibrations per second. The long section of the ruler vibrated slower than this rate. Therefore you heard only

the rapping noises against the table. Above this frequency you began to hear a musical tone.

Why does the pitch go up when you tighten the string on your milk carton instrument? When you pull the string to one side and let go it springs back faster if the string is tighter. Therefore the vibrations are more rapid, the frequency of the sound wave is higher, and the musical tone has a higher pitch.

Transmitting Sound

Here is a simple way to make a fork sound like a bell. Be sure that you use an old fork that can take a few more scratches. Suspend the fork from a thread and strike it with another old fork held in you hand. The fork on the thread gives out a sound like a bell or chimes. The sound lasts for quite a while.

You continue to hear the sound because the fork is free to vibrate and continues to push the air to make a sound wave. But you don't hear much sound from the fork held in your hand because your fingers grip it and tend to stop its vibrations.

Touch the finger holding the thread to the inside of your ear. The sound is now very loud. Why?

The string is stretched by the weight of the fork. The vibrating fork

Fasten the ends of a long string to the bottoms of two cans. Stretch the string and speak into one can. The sound is transmitted through the "string telephone" and is heard at the other end.

Make a speaking tube from a long garden hose. The sound is transmitted from one end to the other by air trapped in the tube.

Make a megaphone from a large piece of cardboard. It makes sounds louder by concentrating the vibrations in one direction. Use it in reverse as an "ear trumpet" to concentrate vibrations toward your ear.

pulls the string and causes it to vibrate. When the bottom of the string is pulled it causes the top of the string to be pulled. In this way the vibration is **transmitted** (sent) up to your ear through the string and you hear the sound.

Hold an old fork in your hand and strike it against a hard object. Be sure that you don't pick the best table in the house. Use an old block of wood or the heel of your shoe. After striking the fork place it on the table. A ringing sound is heard, and continues to be heard.

Strike the fork again and touch only the bottom of the fork to the table top. A loud ring is heard. Lift the fork off the table. The ringing sound almost disappears. The louder sound is heard because the vibrations of the fork make the table vibrate, too. Then the entire table top vibrates and causes more air to vibrate, thus making a louder sound.

Repeat the experiment, this time with your ear to the table while the vibrating fork is held at arm's length, touching the table. You hear a loud ringing sound in the table. The vibrations of the fork are transmitted through the table to your ear.

We normally hear sounds transmitted through air. But in fact any material will transmit sound.

Turn on the radio in a room with closed doors while you listen with your ear to the wall in the next room. The sound is transmitted through the solid wall.

Place different objects between your ear and the wall, such as a block of wood, a pillow, a glass. Solid objects, such as wood, transmit the vibrations much better than loose and fluffy materials, like pillows, because they are hard and rigid, and vibrations can pass through them more easily. When we want to soundproof a room we absorb the sounds by using soft drapes, carpets, and soft ceiling materials.

Why did your milk carton instrument sound so loud? The vibrations of the string were transmitted to all parts of the box. The sides of the box then set the air into vibration. Thus a larger amount of air was made to vibrate than the string itself could push. The sound was there-

fore much louder with the box. That is why violins and guitars have hollow boxes.

Does water transmit sound? The next time you go swimming try banging two rocks in the water while your head is under water. You hear the sound. The vibration is transmitted through the water.

Suppose that an H-bomb was sent up in a rocket to the moon to explode on its surface. We could see the explosion. Could we hear it? No, because there is no air out in space to transmit the vibrations to our ears. It would be a silent explosion!

Speed of Sound

The next time you watch a baseball game try this experiment. Walk several hundred feet away from home plate. Watch and listen as the batter hits a ball. First you see the bat hit the ball. A moment later you hear the sound.

Or, watch a fast airplane as it zooms through the air high in the sky. The sound will seem to come from a point in the sky far behind

SOUND SEEMS TO COME FROM HERE

where you see the airplane. This happens because sound takes some time to reach you.

Light travels so fast (186,000 miles a second!) that you see something happen on earth at practically the instant it happens. But sound travels much slower, about 1100 feet in a second.

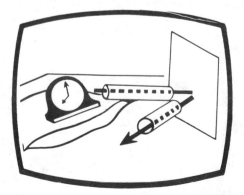

Reflect a sound wave. Place two long wide tubes on the table and place a clock on a pillow at one end. When a cardboard is placed near the ends of the tubes it reflects the vibrations to your ear and you hear the sound.

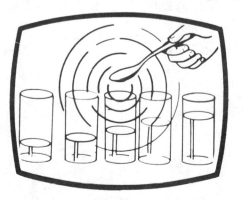

Make the tones of the music scale by tapping a set of glasses. Adjust the pitch of the sound by adding water to each glass to change its rate of vibration.

Blow across the top of a small vial. The air in the vial is set into rapid vibration and makes a high pitched tone. A taller vial or bottle gives a lower tone. An organ makes music in a similar way.

SOUND

Many airplanes travel faster than this.

If you are 220 feet away from a batter as he hits the ball, you hear the sound about 1/5 of a second after he hits it. This difference is easily noticed if you observe carefully. If an airplane is 3 miles up it takes about 15 seconds before you hear the sound.

You can figure out how far away a distant lightning flash occurs by measuring the time for its sound (thunder) to reach you. Sound travels a mile in about 5 seconds. If you hear the sound of thunder 10 seconds after the flash, then the lightning occurred 2 miles away.

Echoes

Try this experiment. Stand near the middle of a flat wall of a large house or barn. Walk directly away from it for about 50 normal steps. Turn around and shout at the wall. You hear an **echo** shortly after your shout stops.

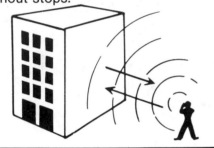

Try this experiment at a greater distance from the wall. Now the echo takes a longer time to reach your ears.

When you make water waves in a bathtub or sink you can see them reflect (bounce) off the sides and return to the starting point. Sound waves can also be reflected. The reflection of your voice is called an **echo.** You can hear echoes from almost any large surface such as a cliff or large building. The further you are from the surface the longer the time needed for the sound wave to reach it and bounce back to your ear, and the greater the delay between the time you shout and its echo.

How We Speak

Touch the front part of your throat while you speak. You feel vibrations as long as you continue to speak. Inside your windpipe is a "voice box"

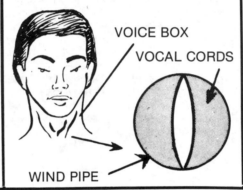

VOICE BOX

VOCAL CORDS

WIND PIPE

or **larynx.** Two flexible **vocal cords** are stretched across this box. When the vocal cords are stretched tightly they vibrate more rapidly, like the string on your milk carton, and thus produce a higher pitch. The air blowing through your throat sets the vocal cords vibrating. The sound is then carried up to your mouth and out.

Your brain sends messages to the vocal cords and they tighten or loosen and therefore vibrate in different ways to create many different sounds.

Try making sounds like: ah, oh, ee, sh, th, br, jump, etc. Sing a very high note and then a very low one. Notice how each sound is produced by different shapes of the lips and tongue along with different sounds from the vocal cords.

The Importance of Sound

The ability to **communicate** ideas to each other, and the ability to use our hands are the two main differences that set us apart from animals and enable us to do so many different things.

We communicate mainly by speaking. And speaking led to writing, to books and newspapers. Children can then learn at school the knowledge gathered slowly by people through many thousands of years. Thus, our modern civilization was built up to what it is today mainly as the result of our ability to make and hear many different kinds of sounds.

TRY THESE EXPERIMENTS

1. Blow between two pieces of paper stretched between your hands. The moving air sets the paper into rapid vibration and creates a high-pitched sound.

2. Put your head inside a pail and sing. The sound is very loud. The metal and air in the pail are set into vibration to reinforce the original sound.

3. Fill a large jug with water. Note the rising pitch of the sound as the length of air in the bottle becomes less. Note how the pitch changes as water is poured out of the bottle.

SCIENCE PROJECT

Open up the boards covering the strings of a piano. Step on the loud pedal to release the felt pads that touch the strings and muffle the sound. Sing a loud steady tone into the piano. Stop and listen. Do you hear the same tone coming out of the piano?

The vibrations from your voice have set those strings into motion which vibrate at the same frequency as your voice.

Sing a different tone into the piano. Try a rapid sequence of tones. Do you always hear the same tones that you sing?

Try this with musical instruments other than the piano.

LIST OF SCIENCE PROJECTS • Based on Experiments in this Book

Plants
Refer to pages 5-8

1. Can you grow a plant upside down?
2. How do vines grow?
3. Grow a plant under artificial light.
4. How does a plant grow from a seed?
5. What conditions are needed by a plant to grow?
6. What is the best amount of water for plants?
7. What is the best amount of light for plants?
8. What is the best proportion of minerals for plants?
9. What minerals are needed for plants to grow?
10. How much weight can a growing seedling lift?
11. Can plants be overcrowded? What is the best spacing for plants?
12. Do plants need carbon dioxide?
13. How does water get to the leaves from the roots?
14. Can fish live without plants in the water?
15. How do molds grow?
16. What uses are made of plants?
17. Can plants grow from the "eyes" of a potato?
18. How can plant foods purchased in a fruit and vegetable store be planted so as to grow into new plants?
19. What plant foods contain starch?
20. What different kinds of plants grow from grass seed?
21. Investigate the growth of algae in a fish bowl or in a pond.
22. Do seeds develop into plants when it is cold?
23. Are plants harmed by heat?
24. Investigate the action of tendrils.
25. What effect does gibberelin have on a plant?
26. Do plants grow equally well in light of different color?
27. Can plants grow from leaves?

Your Senses
Refer to pages 9-12

1. Demonstrate illusions involving the sense of sight, touch, hearing.
2. Investigate the location of spots on the skin that are sensitive to pressure, pain, heat, cold.
3. Investigate the spacing of nerve endings in the skin in different parts of the body.
4. Investigate after-images produced when you stare at at bright scene or picture and then look up against a light background.
5. How helpful are two eyes in judging distances of objects, as compared with one?
6. Make up a set of boxes with objects of different number and shape. Test people as to their ability to determine what objects are in each box from the sounds as the box is shaken.

7. To what extent can different people judge the direction from which a sound comes?
8. How do people differ in ability to match the pitch of two sounds? Use a tape recorder to record piano tones. For each sound ask the subject to pick out the tone on the piano. Keep a record of mistakes. (Do this with a wide variety of tones.)
9. How sensitive are people to smells? Pour some perfume, alcohol, ammonia, or other liquid with an odor, into a hidden, open container. How close do people have to come to the container to recognize the odor?
10. How important is the sense of smell in tasting foods?
11. Do people get used to odors?
12. Do different people differ in their ability to tell which of two very thin objects is thicker?
13. Investigate the location of areas on the tongue that detect different tastes.
14. Do people differ in ability to taste? Make up a series of solutions of sugar, salt, or other food materials of different dilutions. Can some people taste very dilute solutions that others can't detect?
15. Investigate the "blind spot." Do different people reveal a "blind spot" at the same distance from the test patterns?
16. To what extent can a person tell what a substance is by feeling its surface?
17. Investigate the way the pupil of the eye reacts to different brightness of light.
18. Do people differ in ability to hear sound of very high pitch?

Water
Refer to pages 13-16

1. Make different arrangements of water wheels that could be used to obtain motion from moving water.
2. How much of a drop in temperature can be produced by evaporating water? Do other liquids produce more or less cooling?
3. Investigate the ways in which bubbles in soda water and bubbles in regular water are alike or different. Do they cling to the sides of the glass in the same way? Are they the same size and shape?
4. On a cold day water from the faucet sometimes appears cloudy because of bubbles. This cloudiness disappears in a few minutes as the bubbles form and move out into the air. Observe such cloudy water. Can you see the bubbles? Are they driven out more quickly by heat or by cooling?
   Devise a way to estimate how much air comes out in the form of bubbles.

5. Investigate the way running water washes away sand, soil or gravel.

6. Use water to measure the amount of empty space in a jar of marbles.

7. Can water dissolve more salt or sugar? Which solids dissolve best in water?

8. What effect does heat have on the ability of water to dissolve different substances?

9. Does tap water have dissolved substances in it? Does water from one town differ from water obtained in another town?

10. How does water get to the leaves from the roots of a plant?

11. How does the cooling ability of ice compare with that of ice water?

12. Compare the amount of heat needed to bring water to the boiling temperature with the amount needed to boil it off completely.

13. Does it take more or less heat to heat water than to heat solids such as steel or rock?

14. Show that there is water vapor in the air.

15. Show that water is formed when wood burns. Is water formed when other substances burn?

16. Can you make small, artificial clouds with cold materials such as ice or dry ice?

17. How much water does soil contain?

18. What fraction of the weight of an apple is water? Does this differ for various foods?

19. Electricity from a flashlight battery can be used to separate water into the two elements of which it is composed—oxygen and hydrogen. Show how to do this. Can you provide evidence that water is H_2O?

Surface Tension   Refer to pages 17-20

1. What kinds of screens can be made to stay on top of water? Try plastic, aluminum, steel and cheesecloth screens.

2. Does a piece of screening remain on top of the water if the water has detergent in it? How much detergent makes a difference—one drop in a glass; two; three; etc?

3. Does a piece of screening stay up as well on a concentrated salt solution, as on plain water? What happens with sugar and other substances that dissolve in water?

4. Fill a soda bottle with water—almost to the brim. Use a dropper to make the upper surface very flat. How many drops of water can you add to the top before it spills over? Is the amount the same for a plastic bottle as for glass?

5. Water in a glass rises slightly all along the edge. Notice the curve that the water makes with the glass. Compare the curves for a clear plastic cup and a glass. Compare them for an oily glass and a clean one.

6. Compare the cleansing action of soaps, detergents and other cleaning agents. Cut pieces of equal size of cloth and make them dirty with oil and mud. Place each in a dish with a cleansing solution. Let them stand for a period of time. Remove and rinse the cloths. Which is cleaned most thoroughly?

7. Place a drop of water on wax paper. Note the shape of the drop from the side. Is the shape of the drop the same for oil, salt water, sugar solution, other liquids? Is the shape of the drop the same when placed on metal, wood, paper, stone, plastic?
Your observations can be developed into a way to compare surface tension. If surface tension is less, the drop is flatter on wax paper.

8. Pour a small amount of cooking oil on water in a wide pan. Mix the oil and water. Investigate the way the oil forms drops that gather together. Observe what happens with small amounts of soap or detergent in the water.

9. Slowly add alcohol to water in a glass. It will remain on top of the water. Add a quantity of oil. A large round ball of oil forms along the boundary between the alcohol and water. How large a drop can you make this way. Is it perfectly round or flattened?

10. What types of liquids or solutions make the best bubbles?

11. Dip a small metal hoop into bubble solution. It forms a flat bubble surface. Try this with different wire shapes: round, square, triangle. Try it with a wire cube, and more complex shapes. Investigate the way bubble surfaces curve on some wire forms.

12. Some types of model-airplane cement, when placed on a matchstick floating on water, can cause it to move about on the water. Some types of camphor and moth-repellent crystals can also do this. Try different substances and find out which cause such motions and why. Caution airplane glue should be kept away from eyes and nose.

Air Pressure   Refer to pages 21-24

1. Devise a way to measure the force of moving air against objects of different shape and size. Find out why this force is important in the operation of automobiles and airplanes.

2. Devise different methods for measuring the weight of air.

3. Cover a full glass of water with a thin sheet of plastic or metal and invert it over a sink. Why doesn't the water fall out?

   Does the water stay up if there is a small hole in the sheet? A large one?

   Does the water stay up if the glass is not completely full? Half full? One quarter full? Find out why.

   Investigate this effect when the glass is covered with cheese cloth, metal screening, paper or plastic.

4. Do suction cups stick equally well to different surfaces? Try surfaces that are: a. wet, b. dry, c. oily, d. smooth, e. bumpy.

5. Does a suction cup stick to a smooth surface that has a tiny hole?

6. Examine the action of a suction cup by viewing it through a transparent piece of solid plastic as you try to pull away the suction cup. It will be helpful to moisten the surface with water or oil.

7. Invert a long vial in a very tall cylinder filled with water. Slowly push the vial down under the water. Observe the way the air in the vial is compressed. Measure the amount of compression. How is it related to depth?

8. Investigate the way a container filled with water empties if there is only one small hole for the water to pour out. What happens when there are two holes.

9. Crush a can with air pressure. One way is to boil water in the can, cork it, then cool the can by pouring water on it. Another way is to pump out the air. Do round cans crush as easily as those with square corners?

10. Observe the changes in a barometer from day to day. How accurately can you predict the weather from the changes?

11. Carry a barometer up and down in an elevator of a tall building. How much does the barometer change for a rise of 50 feet? 100 feet?

    with an air hammer. Find out how this device works.

13. Operate a siphon. How is the rate of flow affected by the width of the tube and the difference in the water levels?

14. Blow air from a vacuum cleaner at some ping pong balls and balloons. Make them stay up in the air. Find out why they stay up.

15. Investigate the way we breathe.

Carbon Dioxide
Refer to pages 25-28

1. Investigate the way foam is formed when dry ice is placed in water with detergent. Test the effect of the following: a. different detergents, b. hot or cold water, c. wide or narrow container, d. temperature of the air, e. different amounts of detergent in water.

2. Show that carbon dioxide is heavier than air.

3. Which has greater ability to cool water, dry ice or regular ice?

4. Show that air has a small amount of carbon dioxide.

5. Show that breath has more carbon dioxide than air.

6. Show that carbon dioxide is formed when vinegar is added to bicarbonate of soda.

7. Can other substances besides vinegar and bicarbonate of soda produce carbon dioxide? Try the following instead of vinegar: lemon juice, citric acid, other acids, salt solutions.

   Try the following solids instead of bicarbonate of soda; sodium carbonate, potassium carbonate, other carbonates, baking powder, limestone, chalk, other types of rocks.

8. Investigate the conditions which cause baking powder to release bubbles of carbon dioxide when bread is baked.

9. Prove that a burning match produces carbon dioxide.

10. Under what conditions does soda water retain its carbon dioxide. Test the following factors: a. temperature of surroundings, b. closed or open container, c. temperature of soda water.

11. How rapidly does dry ice evaporate under different temperature conditions? Test it in the refrigerator, on the table, and in a stove at different temperatures.

12. A chemical known as bromthymol blue can be used to test the action of water plants on carbon dioxide in water. Obtain some of the solution. Blow through a tube into the solution, thereby adding carbon dioxide. The solution turns yellow. Put in a small water plant of the kind used in fish tanks. Place the plant in sunlight. After a while the solution turns blue again because carbon dioxide is used up by the plant in sunlight.

    Experiment with this test to find the effect of the following: more or less carbon dioxide, more or less sunlight, different temperatures, and other factors.

13. Carbon dioxide is one of a number of common gases—oxygen, hydrogen, nitrogen, methane, helium, etc. Find out how these gases are made, and their properties.

Bicycles
Refer to pages 29-32

1. How effective are wheels in reducing friction? Use a spring scale to measure the force required to pull a skate on its wheels, and again when being dragged on its side. How do these forces compare?

2. How much difference in friction force is there when objects are dragged on smooth surfaces and rough ones?

3. How much effect do roller bearings have in reducing friction in a wheel?

4. Devise a way to measure the force required to pull a boy on a sled on concrete. Compare this with the force required when he is on a bicycle. How do you explain any differences observed?

5. Compare the forces required to pull a boy on a bicycle on different types of surfaces - asphalt, cement, grass, rocky ground, mud, etc.

6. Count the number of teeth in the sprocket and rear gear of a bicycle. Measure the diameter of the wheel. From this information predict how far a bicycle will move for one turn of the pedals.

7. Compare a girl's bike with a boy's bike as to the distance moved with one turn of the pedal. Do the same for bicycles with gears. Which gear gives the greatest distance moved for one turn of the pedals? Find out why these differences occur.

8. When a bicycle is motionless it topples very easily if support is removed. Yet when moving it has a strong tendency to keep erect. This is related to the principle of the gyroscope. Investigate these effects with a toy gyroscope.

9. Use a bicycle to measure the distance between two places.

10. Analyze the arrangement of spokes which connect the hub of a bicycle to its outer wheel. How do these thin wires make the wheel rigid?

11. Analyze the motions a rider makes as he rounds a turn. Does it make any difference how the rider leans over?

12. Old-fashioned bicycles had very large front wheels turned directly by means of pedals, and a small rear wheel. Why were these bicycles made that way? Why is the modern form of bicycle easier to ride? What parts do the sprocket, chain, and gear play in improving the operation of the bicycle?

13. Increased friction is important in the operation of bicycles, mainly with regard to the way the wheels grip the road. Measure the friction force when a locked bicycle is dragged on its wheels. Is the friction force greater or less when the tire is worn smooth than when it is new?

14. Compare different bicycles as to width of the tires. Which require more air pressure in the tires? Why? What are the advantages and disadvantages of thick tires versus thin ones?

15. Find out how the braking system of a bicycle works.

Flying
Refer to pages 33-36

1. Demonstrate Bernoulli's Principle with: a. card and book, b. folded paper, c. vacuum cleaner and ping pong balls, d. spool and card, e. strip of paper, f. two apples suspended from strings.

2. Investigate the ability of an upward stream of air from a vacuum cleaner to lift ping pong balls, balloons, styrofoam balls, wooden balls, and other types of balls.
What effect does each of the following factors have: a. diameter, b. density, c. roughness of surface, d. velocity of air stream e. width of air stream, f. turbulence of air stream, g. shape?

3. A small ball can be lifted into the air and kept in one position by means of an upward jet of water. Investigate the following factors regarding this effect: a. weight of ball, b. velocity of jet, c. material, d. type of surface.

4. Blow toward a soda bottle and extinguish a candle behind the bottle. Does this work with smaller or larger bottles? Does it matter if you blow close to the bottle or farther away?

5. Make models of parachutes and investigate different designs that lower a falling weight at a slow speed. Which are most effective?

6. Make models of different shapes and measure their drag in an airstream.

7. Measure the force of wind against a large board.

8. In 1967, *Scientific American* magazine awarded prizes in a contest for designs of paper airplanes that would remain in flight for the longest possible time. Design your own models which stay aloft as long as possible.

9. Find out how the control mechanisms of an airplane—rudder, elevator, ailerons—enable the pilot to maneuver.

10. How does a pilot control the flight of an airplane?

11. Investigate the principles of flight of a helicopter.

Earth Satellites
Refer to pages 37-40

1. Investigate methods of making a rubber balloon go as high or as far as possible by means of reaction from air rushing out of its open end.

2. Investigate the amount of reaction to the spin of a propeller in a model airplane.

3. Investigate the centrifugal force produced on objects placed on a phonograph turntable. Try the effect of different speeds.

4. Demonstrate inertia for objects at rest, and for objects in motion.

5. Float a metal bowl on water in the sink. Add objects of various types—large solids of different shapes, small particles (sand or salt), or liquids. Investigate the way the bowl rotates in the water with these different objects inside. What principles of motion are shown?

6. Show that a lawn sprinkler works on the same principle of action and reaction as a rocket engine.

7. Build a reaction engine that operates by means of falling water. Use a rectangular plastic bottle. Drill holes in opposite sides near the left bottom corner. Suspend the bottle from a string with a loop centered over the opening. Hold the bottle over a pan or sink and fill it with water. As the water pours out, off-center, the actions cause reactions that make the bottle spin. Experiment with different sizes of holes to see which produce the greatest speeds or number of times of rotation.

8. Prove that a gas expands when heated.

9. Find out how a gas turbine engine works.

10. Investigate (by means of library research) the operation of different types of jet and rocket engines.

11. Make a working model of a reaction-driven boat. You might use an inflated balloon for propulsion. Or, you might use bicarbonate of soda and vinegar in a small plastic squeeze bottle to generate carbon dioxide gas for power.

12. Make a model of a space station. Investigate (by reading books) the problems men would encounter in setting up a community on the moon or in space.

13. The action-reaction principle is used to steer a space ship and control its motion. Investigate how this is done.

Center of Gravity   Refer to pages 41-44

1. Use a weight on a string to find the center of gravity of a flat object of any shape. Can the object be balanced in a horizontal position on a nail head placed under that point?

2. Place a flat object on the table and slowly push it toward the edge until it falls off. Do this with the object in several different positions. Can you use the information obtained in this way to locate the center of gravity of the object?

3. Find the center of gravity of a three dimensional object such as a chair.

4. Show that an object (such as a toy car) is more stable if it has a lower center of gravity and a wider base.

5. Investigate and explain the stability, or lack of it, for objects of various shapes, such as balls, cones, blocks.

6. FInd a way to predict the exact position from which a tall block of wood will topple over when tipped.

7. Investigate the part that center of gravity plays in balance when a person walks, and when he carries heavy objects.

8. Find out how to locate the center of gravity of flat objects shaped like rectangles and triangles by drawing straight lines.

9. Find out how to predict the location of the center of gravity of any polygon (flat shape with straight sides).

10. Investigate the way people standing in a bus increase their stability as the bus moves.

11. If the center of gravity of an airplane is improperly placed, the airplane will be unstable in flight and tend to stall and crash. Placement of cargo and people in the airplane affects the position of the center of gravity, and so may affect the flight of the airplane. Find out how this problem is solved in airplanes.

12. The bottom surface of a man's foot is much larger than that of the hoof of a horse. Investigate the reasons for this and the part that the center of gravity plays in determining the size of the foot of an animal.

Magnetism   Refer to pages 45-48

1. Test a number of magnets and find out under what conditions they attract or repel.

2. Make a compass from a suspended magnet. Make one from a magnet floating on a liquid.

3. Use iron filings to show the lines of force around a magnet. How do the lines of force appear when three magnets are near each other? Four magnets?

4. Make a complex structure of iron objects that is held together by magnets.

5. Through what materials will magnetism pass?

6. Test the strength of various magnets.

7. What materials does a magnet attract?

8. Find the strongest parts of different magnets.

9. How can you make a magnet?

10. What happens if you break a magnet?

11. Make magnets from various steel and iron articles. Do they retain their magnetism?

12. Show how to detect a weak magnet by using a compass.

13. Magnetize a steel needle and then show how its magnetism can be destroyed by heat.

14. Show that a magnetic field is created in a wire by electric current. Make a chart showing the direction of the current and the magnetic field.

15. Show that electric current flowing through a coil of wire can magnetize iron.

16. Make an electric current detector from a compass and a coil of wire around it.

17. Make a working model of an electric bell or buzzer. Change it into an electric gong by means of a change in the wiring.

18. Show that pushing a magnet into a coil of wire can generate electric current. How is the direction of current affected by the position of the magnet and its poles? How can the amount of current be increased?

19. Show how to make an electromagnet from a coil of wire wrapped around an iron nail.
20. Make a model of an electromagnetic crane and explain how it operates.
21. Make a model of a telegraph key and receiver and explain how it works.
22. Test various stationary iron objects in your home to see if they have magnetic poles. Where are these poles located?
23. Compare the strengths of magnets made of steel, of alnico, and of other magnetic materials.
24. Investigate the magnetic fields of the ends of bar or rod magnets.
25. The magnetic field of a strong magnet can wipe out the magnetic recording on a tape recording. At what distances from the tape do different types of magnets begin to affect the recording?

10. Charge a plastic comb or sheet by rubbing it. Place it near some bits of paper, string and other small objects so that they cling to it. Observe the motions of the small object. Explain these motions and any differences you observe between various materials.
11. Charge a plastic dish cover by rubbing it. Place some bits of paper, threads, and other small objects in the cover and move you fingers underneath. What motion do you observe? Explain these motions.
12. Use static electricity to separate a mixture of salt and pepper. What other mixtures of powders can you separate in this way?
13. Investigate the abiltiy of charged objects to cling to a wall. How does the weather affect this abiltiy? Which materials stay up the longest? Does the weight or shape of the material affect the time it stays on the wall?

Static Electricity   Refer to pages 49-52

1. Investigate different types of materials to find out which produce static electricity when rubbed on other materials. Which combinations are best?
2. Investigate sparks produced by static electricity in a very dark room.
3. What effect do cool, warm, humid, or dry weather have on the production of static electricity?
4. Produce light in a flourescent lamp by rubbing it with nylon cloth. Do this with other types of materials. What types of cloth are best? How can you increase the amount of light produced?
5. A rubber balloon takes on a (—) charge when rubbed on cloth. Use this fact to identify the type of electric charge produced on various plastics.
6. When two different materials are rubbed and a (—) charge is produced on one, is a similar charge produced on the other?
7. Investigate the ability of various charged materials to attract small objects such as: a. bits of paper, b. pieces of thread, c. straw, d. puffed cereals, e. feathers.
8. Enclose bits of paper, thread, and other small objects in a transparent plastic container. Rub the top of the container with various materials (nylon, cotton cloth, plastic sheets). Investigate the motion of the small objects. Does the size, shape, or material of the container make any difference?
9. Sparks may be produced by shuffling across a rug and touching a metal object such as a radiator or metal base of a lamp. Do rugs differ in the ability to produce sparks? Does the weather matter? Does the method of shuffling or type of shoe bottom make a difference?

Electric Current   Refer to pages 53-56

1. What arrangement of connections among several flashlight cells produces the loudest sound in an electric bell? Use a voltmeter to measure the voltage produced by each combination.
2. Make and operate a rain alarm.
3. Make and operate a burglar alarm.
4. Make and operate a fire alarm.
5. Electricity from a flashlight cell, passed through water to which some vinegar has been added, decomposes the water into oxygen and hydrogen. Devise a way to collect and identify the gases. Are they formed in equal amounts?
6. Use a flashlight cell to produce sparks when a wire is rubbed on a rough surface such as a file. The sparks may be made brighter by connecting the wire to a "transformer". Observe the sparks and find out what causes them to be produced.
7. Sparks may be produced by momentarily short-circuiting a flashlight cell. These sparks create radio waves that can be detected by a nearby radio set as "static". Investigate the properties of this type of simple radio transmitter. Does an aerial on the transmitter help? Does the position or shape of the aerial make any difference? Do the radio waves go through wood, solid metal, plastic, solid walls, water?
8. Investigate the methods of connecting switches so as to light several lamps in different ways.
9. Investigate the operation of fuses.
10. Test various metals and other substances for ability to conduct electricity.
11. Find out what is meant by electrical *resistance* and how it can be measured with an ammeter and voltmeter.

12. Batteries are sold that produce much larger voltages than those produced by one cell. Open up used batteries of this type and find out how they produce higher voltages.
13. Make an electromagnet. Use it to pick up small iron objects such as clips, nails and washers. Find out how to make these electromagnets stronger.
14. Make a working model of a telegraph set.
15. Make a working model of a buzzer.
16. Make a working model of an electric motor.
17. Find out how to plate metal objects with copper, silver, nickel and other metals. Plate various metal objects with different metals.
18. Investigate the differences between series and parallel circuits.

Light and Sight
Refer to pages 57-60

1. Under what conditions can glass be invisible?
2. How can water be used to see around a corner?
3. Investigate the effects of water in a glass on the appearance of objects in it, or beyond it.
4. Can clear marbles form images?
5. Investigate imperfections in windows by the distortions of appearance of objects seen through the window.
6. Investigate refraction, the bending of light rays by transparent materials.
7. Investigate the ability of a lens to set fire to a small piece of newspaper. Be sure to take precautions against fire.
8. Investigate the manner in which lenses form images.
9. Investigate the part lenses play in a camera.
10. Find out how the lens in the eye enables us to see.
11. Investigate the ways in which the eyes of insects and crustaceans (lobsters, etc.) differ from ours.
12. How does a magnifying glass produce enlarged images?
13. Investigate the optical system of a projector.
14. Investigate the action of concave lenses on light.
15. Find out why light bends.
16. Find out about the wave theory of light, and how it can account for reflection and refraction.
17. Investigate mirages by means of photographs. A simple type of mirage is often seen on a hot day while riding in a car. A "pool" of water—a mirage—is often seen in the road ahead of you.
18. Investigate how lenses in eyeglasses can improve vision.
19. Investigate the dancing patterns of light seen on the bottom of a shallow pond on a sunny day.

20. Investigate the light patterns on the bottom of a shallow pool on a sunny day, produced by small objects resting on the water surface.
21. Use photographs to investigate the way in which the sun appears to become flattened as it nears the horizon.
22. Investigate how lenses are used telescopes.

Mirrors
Refer to pages 61-64

1. Make a display showing how mirrors are used.
2. Make and demonstrate a kaleidoscope.
3. Arrange a set of mirrors that show people how others see them.
4. Investigate the appearance in a mirror of printed words. Learn to write so that it appears normal when viewed in a mirror.
5. Investigate the difficulty people have in writing or tracing lines while viewing the writing hand in a mirror.
6. Make a periscope and use it to see around a corner.
7. What is the difference between regular and diffuse reflection? Why are they important?
8. Show the difference between: transparent, translucent and opaque.
9. Produce the illusion of a mirror roadway by placing two mirrors near each other face to face.
10. Produce multiple images with two mirrors and show why they are formed.
11. Make a hall of mirrors with three mirrors.
12. Investigate how a curved mirror forms an image.
13. Produce an illusion in which a candle appears to be burning in a jar of water.
14. Photograph reflections in curved surfaces (car fenders, spoons) and analyze the nature of the images and distortions.
15. Investigate the images formed by a concave mirror of the type used for shaving.
16. Photograph and study the reflections in a fish tank.

Heat
Refer to pages 65-68

1. Do all liquids expand when heated? (Caution: do not heat flammable liquids.)
2. An exception to expansion with heat occurs when ice is heated and melts. It then contracts. It also contracts as the melted water at freezing temperature is heated from 32°F. (0°C) to 39°F (4°C). Investigate these exceptions.
3. Doing work of any kind always causes heat to be produced. Show that this is so by bending, twisting, rubbing, or hammering objects such as metal wires and sheets.

4. Do different gases expand the same amount for a given rise in temperature, or by different amounts?

5. Devise a way to measure the amount of expansion of a metal tube heated by passing steam through it. Exposed steam pipes in basements are convenient for such investigations because their length produces greater expansion than for short tubes.

6. How accurately do people judge temperatures?

7. How accurately do different thermometers measure temperatures?

8. Demonstrate how heat can be transferred by means of convection currents of air.

9. Demonstrate how heat can be transferred by convection currents in water.

10. Investigate the direction of winds near the shore of a large lake or ocean. Do the winds blow differently by day or by night?

11. Show that different metals conduct heat differently.

12. Investigate our judgements of temperature of various materials (metals, wood, glass, etc.) when we touch them with the fingers on a cold day.

13. Try heating water by radiation from a lamp. How can you increase the amount of heat absorbed?

14. Investigate the ability of a lens to focus sunlight and set a piece of paper afire. Does the diameter of the lens make any difference? Does the color of the paper matter?

15. Make a model of a house heating system that uses steam, hot water, or hot air.

16. Show that evaporation of a liquid causes cooling.

17. Make a candle out of butter. What other soft solids and waxes can be made into candles?

18. The amount of heat given up by a hot object can be measured from the rise in temperature produced in a given quantity of water. Which substances are most effective in this respect?

19. Measure the caloric value of different foods.

Sound
Refer to pages 69-72

1. Make musical instruments from simple materials such as: a. box and rubber bands, b. box and nylon threads, c. metal tub and steel strings, d. comb, e. glasses containing water, f. vibrating steel strips, g. hollow tubes.

2. Investigate waves in a pond by throwing stones into the water. What shape do the waves have? What happens when two waves meet? How fast do water waves travel? Does the speed depend upon the size of the wave?

3. Make a stethoscope and investigate the sound of heartbeats.

4. Open the sounding board of a piano and step on the loud pedal to release the felt pads that muffle the sound. Pluck the strings and investigate the way sounds of different pitch are produced in the piano.

5. How far can you transmit speech through a string with a string telephone? What types of string are best?

6. Make a string telephone from a long length of wire. Keep the wire stretched. Listen to it when the wind blows. Can you tell from the sounds you hear at the end of the wire how fast the wind is blowing?

7. For how long a distance can you transmit speech through a tube? Does the tube have to be straight?

8. Make megaphones of different shapes and sizes. Which are best for transmitting sound for greater distances?

9. Investigate the way in which sounds travel through different types of material—solid walls, tables, soft materials, steel.

10. Investigate the ability of water to transmit sound. Is it better for this purpose than air or steel?

11. Measure the speed of sound by noting the time required for a sound to travel a large distance. A distant steam whistle on a boat or at a factory is excellent for this purpose because the puff of steam is often visible for several miles.

12. Measure the speed of sound by means of echoes from a distant cliff or large building.

13. Analyze the way different sounds are produced by the human voice.

17 Additional Projects. See pages 8, 12, 16, 20, 24, 28, 32, 36, 40, 44, 48, 52, 56, 60, 64, 68, 72.